21

Paramarine!

Uniforms and Equipment of Marine Corps Parachute Units in World War II

Paramarine!

Uniforms and Equipment of Marine Corps Parachute Units in World War II

Chris Mason

Schiffer Military History
Atglen, PA

Dedication
This book is dedicated to every Sailor and Marine who ever jumped out of a perfectly good airplane.

• • •

In Memoriam
Captain John Denning Mason, USN
1915-2002

Photography by Robert Biondi.

Library of Congress Catalog Number: 2003109830.

Printed in China.
ISBN: 0-7643-1924-8

For the largest selection of fine reference books on this and related subjects, please visit our website – **www.schifferbooks.com** – or call for a free catalog.

We are always looking for people to write books on new and related subjects. If you have an idea for a book, please contact us at the address below.

Published by Schiffer Publishing Ltd.
4880 Lower Valley Road
Atglen, PA 19310
Phone: (610) 593-1777
FAX: (610) 593-2002
E-mail: Info@schifferbooks.com.
Visit our web site at: www.schifferbooks.com
Please write for a free catalog.
This book may be purchased from the publisher.
Please include $3.95 postage.
Try your bookstore first.

In Europe, Schiffer books are distributed by:
Bushwood Books
6 Marksbury Ave.
Kew Gardens
Surrey TW9 4JF
England
Phone: 44 (0)20 8392-8585
FAX: 44 (0)20 8392-9876
E-mail: info@bushwoodbooks.co.uk.
Free postage in the UK. Europe: air mail at cost.
Try your bookstore first.

Foreword

Marine Paratroopers and Airborne Units

Early Marine Corps use of the parachute was limited to flight crews and the aerial delivery of supplies. Although there was dormant interest in the use of the parachute to transport Marines to battle, the use of parachutists was still only a sideshow in comparison with the development of amphibious warfare that was the Marine Corps' goal in the 1920s and 1930s. In 1927, a "mass" jump of twelve Marines was made from a transport aircraft flying over Anacostia in Washington, D.C. Parachutists were further used in 1937 during Fleet Landing Exercises on islands off the California coast.

This changed in 1940 when Secretary of the Navy Frank Knox (among others) saw the German use of parachutists in the Low Countries. Recognizing a potential new weapon, Knox directed Marines to form similar parachute and glider units. It was planned to train a battalion of each Marine regiment as an air landing force, including a company of paratroopers. By April 1941, this evolved to the need for a battalion of paratroopers in each division to conduct reconnaissance, raids, and other independent actions.

Between August 1941 and July 1943, Marines formed the 1st, 2d, 3d, and 4th Parachute Battalions of the Fleet Marine Force. All volunteers underwent the rugged training that characterized the Marine Corps jump program at the Parachute School Detachment, Naval Air Station Lakehurst, New Jersey; Parachute Schools, Camp Gillespie, California and New River, North Carolina. Others attended the U.S. Army-run parachute course at Fort Benning, Georgia, particularly those Marine strategic intelligence personnel assigned the Office of Strategic Services. Airborne training was also conducted for a short time during the war on New Caledonia.

Marine parachute units fought from 1942 until 1943 at Gavatu, Guadalcanal, Vella Levella, Choiseul, and Bougainville. The wartime demands on the Marine Corps saw the parachute battalions used in the Pacific as amphibious assault forces rather than in their intended airborne role. Jump training and plans were made for airborne assaults against Kolombangara, Bougainville, and New Ireland. These operations were never carried out due to a lack of transport aircraft and suitable drop zones. Despite their excellent fighting record, Marine parachute units never made a combat jump. Concentrated at Camp Kiser, New Caledonia, as corps reserve, in April 1943 the separate battalions formed the 1st Parachute Regiment. In December 1943, the Department of the Navy disbanded the Marine Corps parachute effort of some 3,000 Marines and Sailors. By February 1944, the airborne units ceased to exist with the exception of air delivery sections.

After World War II the Corps' interest in the use of parachutists was limited to the existing air delivery platoons. There was a proposal to convert a Marine division into an airborne division, but this concept died as vertical envelopment and air transport by helicopters was pursued. Even so, the Department of Defense directed the Marines to train and equip units as needed for airborne operations in conjunction with the U.S. Army and Air Force. Paratroopers needed for this were trained at U.S. Army or Navy schools. In the late 1950s, a resurgence of interest in paratroopers was seen by their use in two new units: the force reconnaissance companies and the air-naval gunfire liaison companies (ANGLICOs). Force recon trained static line and free fall parachutists for pathfinder and reconnaissance missions. ANGLICO trained as parachutists to provide supporting arms liaison teams to the U.S. Army and allied forces. Both types of Marine airborne units served in Vietnam, where the Corps' first combat jumps were made. In 1976, division reconnaissance battalion trained their deep reconnaissance platoons as paratroops, becoming the most recent members of the Marine airborne fraternity.

These modern units all trace their origins to the Marine paratroopers and units of World War II. These are the men and units that author M. Chris Mason documents in this stunning book. It is a fitting tribute to these pioneers to which any parachute trained Marine and airborne unit today must look for their history and traditions. The Marines have never separated their parachutists from the rest of the Corps, although there have been distinct insignia and uniforms over time. Marines trained by the U.S. Army are awarded the basic parachutist insignia. The Navy-Marine Corps parachute insignia was reinstated in 1963, and since then these coveted "gold wings" have been awarded when Marines and Sailors complete ten jumps while assigned an airborne unit for which they receive jump pay.

Those interested in more information on this subject should see Jon Hoffman's *Silk Chutes and Hard Fighting: U.S. Marine Corps Parachute Units in World War II*, Charles Updegraph's *Special Marine Corps Units of World War II*, and John Johnstone's *United States Marines Parachute Units*. Mr. Mason himself is a veteran naval gunfire officer who trained as a paratrooper at Lakehurst, New Jersey and served with the 2d Air-Naval Gunfire Liaison Company, Fleet Marine Force, providing U.S. Army airborne divisions with naval supporting arms. Airborne – All the Way!

Charles D. Melson
Chief Historian, U.S. Marine Corps

Contents

Acknowledgments

Twenty years and more ago, when many of us discovered our interests in the historical artifacts of World War II, there were few reference books available. Much of what we knew we learned from each other, or gleaned from period articles and photographs and the relics that veterans gave us. Today, many new books are available which we could only dream of decades ago.

For the creation of this book I would like to recognize, first and foremost, two outstanding collectors without whose help this book literally would not have been possible: John Busardo of Virginia and Leif Jacobsen of Colorado. Not only did both gentlemen open up their extraordinary private collections for this book and lend the project their vast knowledge of the subject matter, but both also stood ready literally day and night to help, and contributed hundreds of hours of their time to bring it to fruition. Whatever enjoyment readers get from this reference book is due in large measure to these two leaders of our hobby, and I owe them both a great debt of thanks.

Many other people and institutions also made vital contributions to this book, beginning with the United States Marine Corps. I would like to extend my special thanks to the Chief Historian of the Marine Corps, Major Charles "Chuck" Melson, USMC (Ret) for his collaboration in this book, as well as to Jack Dyer and Lena Kaljot at the Marine Corps Historical Center in Washington, DC for their tireless efforts in locating many of the photographs used. I would also like to thank the staff of the Research Center at Marine Corps University in Quantico, Virginia, for their help in finding source material. My thanks also to the Public Affairs Office at Marine Corps Recruit Depot, Parris Island, and to the staff of the National Archives in Washington DC. I would also like to recognize Colonel Dave Severance, USMC (Retired), the Secretary and Treasurer of the Survivors Association of former Marine Corps and Navy parachutists of World War II, for his valuable input and proofreading of the captions. In addition, I would like to thank Major Ed Wages, USMC (Retired) for his encyclopedic knowledge of Marine Corps insignia and equipment, and his patience and good humor in lending it to me for this book. Thanks also to Doug Bailey, John Burnett, Camillus Cutlery Company Historian Tom Williams, Dudley Cone, Harlan Glenn, Clare Mugno, Eric Queen, Don Perrin, Joe Skutt, Jang Younyoung, and George Petersen of National Capital Historical Sales in Springfield, Virginia, who has always been a great source of knowledge as well as original and museum reproduction Marine Corps artifacts from the 20th Century. Finally, I'd like to thank my wife Yasemin and my daughter Hannah for their patience and encouragement in keeping the project on track despite flooded basements, pneumonia, trips to Afghanistan and other calamities.

Author's Note on Nomenclature

The term "Paramarine" was (and still is) discouraged by some Marines who find the double meaning of "para" as an auxiliary element too ambiguous (as in "paramilitary" meaning "an auxiliary military force," for example). The Marine Corps itself, however, used term extensively in World War II, and the term "Paramarine" has remained fixed in the public imagination. Both formulations are used here interchangeably as terms of honor, as is the generic "paratrooper." Although the Marine Corps seldom used the term paratrooper in print during World War II, and does not often use this word today in referring to parachute-qualified personnel (preferring the term "parachutist" or "parachute-qualified"), the American public during World War II did not make this distinction.

Regarding the uniforms and equipment shown, an effort has been made to use the proper Marine Corps terminology ("field scarf" instead of "necktie," for example) without being enslaved by this convention. The often convoluted nomenclature for a uniform item or piece of equipment – "Coat, utility, HBT, camouflage (parachutist)," for example – was sometimes a source of humor among servicemen during the war. After being introduced in the captions these are subsequently shortened for the sake of readability, and terms like "coat" are sometimes used interchangeably with "jacket" or "tunic," for example, to prevent the captions from becoming repetitious. In addition, some official nomenclature of sixty years ago, such as "parachutist overall," for a jump smock, for example, has become obscure today, and the modern term known to military uniform enthusiasts is used instead. Thus the captions may not satisfy pedantic readers, but this book is intended as a readable guide to the uniforms, equipment and insignia of the Marine airborne units of World War II, and not as an official USMC supply manual.

CHAPTER ONE

Uniforms

Right: The forest green winter service coat of Paramarine Sergeant Harold A. Hoeschen, Company C, 1st Marine Parachute Battalion. Hoeschen was wounded in the Paramarine's first combat action, on the island of Gavutu, and later served Stateside as an instructor. (L. Jacobsen)

Below: Paramarine Harold A. Hoeschen. (USMC)

Left: The back side of the Hoeschen winter service coat. (L. Jacobsen)

Below: A close up of the chest and sleeve of the Hoeschen coat. The shoulder patch was officially designated as the insignia of "1st Marine Amphibious Force, Parachute Units." but is generally known as the "1st Marine Parachute Regiment" patch. (L. Jacobsen)

Inset below: Regulations required Marines to stamp or write their names in ink in the lining of the sleeve near the armpit as seen here on Hoeschen's winter service coat. (L. Jacobsen)

Left: Hoeschen's forest green winter service trousers. (L. Jacobsen)

Below: Detail of the winter service trousers. (L. Jacobsen)

The khaki summer service coat worn by Sergeant Harold Hoeschen. This tunic and the accompanying trousers were given to Hoeschen by other Marines when his seabag was lost. His initials are written in the back seam. (L. Jacobsen)

Hoeschen's khaki summer service trousers. (L. Jacobsen)

Above left: Company C, 1st Marine Parachute Battalion, assembled at NAS Lakehurst (New Jersey) during parachute training. Hoeschen is seated in the front row, fifth from left. (C. Mason)

Above: The rear of Hoeschen's khaki summer service coat. (L. Jacobsen)

Left: The khaki service shirt and field scarf worn with both the green and the khaki service coat by enlisted personnel. The shirt was produced in wool for winter and cotton for summer wear. (L. Jacobsen)

SHARPSHOOTER

Opposite left: Paramarine Corporal (here a Private First Class) Charles C. Clow, Jr., of Minneapolis, Minnesota, wearing the winter service green uniform with overseas cap while home on leave. Clow served with 2nd Marine Parachute Battalion and was later killed in action on Iwo Jima with 27th Marines. (L. Jacobsen)

Opposite right: The forest green winter service coat of Private First Class Warren G. McGuire, 1st Marine Parachute Regiment. (C. Mason)

Right: The back side of the McGuire tunic. (C. Mason)

Above: A close up of the "Parachute Man" patch worn above the cuff on the lower sleeve. This patch is discussed in detail in the wings and insignia chapter. (C. Mason)

Left: A close up of the chest and sleeve of the McGuire tunic. (C. Mason)

Below left: McGuire has written his name in the armpit of his tunic in the prescribed manner. (C. Mason)

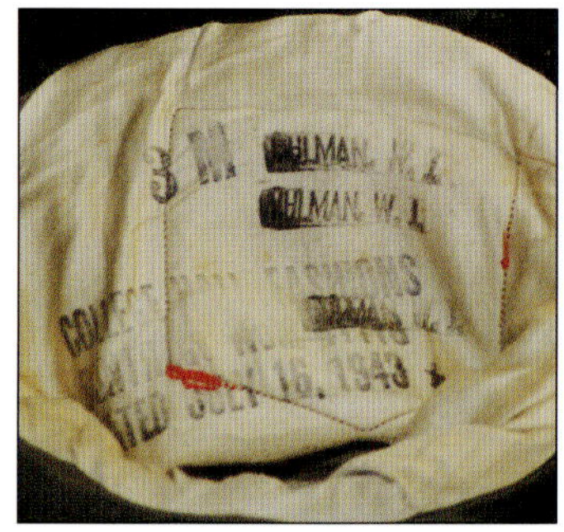

Left: The green winter service coat of Private First Class Walter J. Uhlman, Company C, 1st Marine Parachute Battalion. Uhlman was wounded in action on Iwo Jima with 26th Marines. (L. Jacobsen)

Above: Uhlman has stamped his name in the armpit of his tunic in the regulation manner. Inexpensive kits to make such name stamps were commonly sold on base. (L. Jacobsen)

Right: Two Marines in the winter service uniform. On the right is Paramarine Private First Class (later Corporal) Hermann C. Glass. (E. Queen)

Right: The winter service coat of Private First Class Frederick H. Nicholls, Company B, 1st Marine Parachute Battalion. (L. Jacobsen)

Below: Marine Parachutist Private First Class Frederick Nicholls, wearing his winter service coat with parachutist wings and Paramarine shoulder patch. (L. Jacobsen)

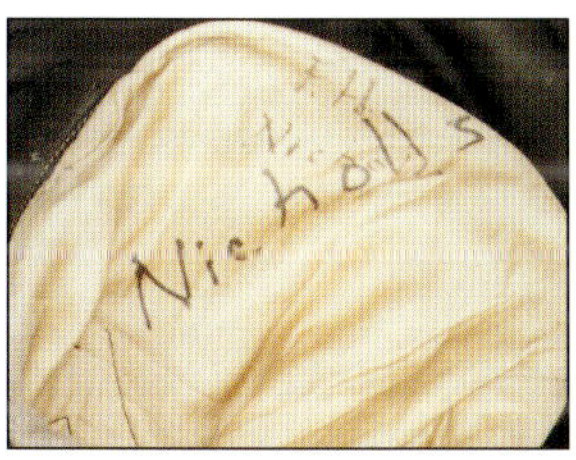

Above: Nicholl's tunic is also named in the armpit area. (L. Jacobsen)

Below: A close up of the Australian-made Paramarine ("1st Marine Amphibious Corps Parachute Units") patch on Nicholl's left sleeve. (L. Jacobsen)

Left: The winter service coat of Colonel Gaylord S. Ziegler, dated 1952. Ziegler served in Headquarters Company, 3rd Marine Parachute Battalion and was later wounded on Iwo Jima with 28th Marines. (L. Jacobsen)

Above: Ziegler as a Lieutenant Colonel wearing his parachutist wings. (L.Jacobsen)

Above: The Australian battledress jacket of Private First Class Edward L. McClure, Company F, 2nd Marine Parachute Battalion. The Australian jacket was widely issued to Marines garrisoned in Australia as a standard uniform item. (C. Mason)

A close up of the chest and sleeve of the McClure jacket, showing the Australian-made Paramarine patch on the left sleeve. (C. Mason)

Left: The tunic is named inside above the manufacturer's label, which is dated 1942, as there is no armpit lining. (C. Mason)

Above: The dress blue uniform coat of Corporal (later Sergeant) Edward G. Gruber. Gruber served in the 1st Marine Parachute Regiment and was wounded on Iwo Jima. Right: The back side of Gruber's Dress Blues. (C. Mason)

Top right: Colonel (later Brigadier General) Robert H. Williams, USMC, wearing dress blues. As a Major, Williams commanded the 1st Marine Parachute Battalion on Gavutu and was wounded there. As a Lieutenant Colonel, he commanded the 1st Marine Parachute Regiment for most of its existence. During the war, Brigadier General Williams earned the Navy Cross, the Silver Star and the Purple Heart, which can be seen on his ribbon bar. (USMC)

Middle right: Corporal Hermann C. Glass wearing the Dress Blue uniform. Glass served with the 1st Marine Parachute Regiment. The Expert Rifleman Badge he is wearing beneath his Basic Parachutist Insignia has a Marine Corps Eagle, Globe and Anchor (EGA) emblem affixed to the point where the rifles cross, an unauthorized but common variation. (E. Queen)

Right: Gruber in Dress Blues after basic training at Parris Island, South Carolina, before volunteering for the paratroops. (C. Mason)

Below: A close up of the chest and sleeve of Gruber's uniform. (C. Mason)

Above: The coat of the special two-piece parachutist uniform of the Marines, designated as a "coat, utility, HBT, camouflage (parachutist)." It was reversible, with a green side, seen here, and a tan or "brown" side. It differs from the standard camouflage utility coat most noticeably in its slanted pockets. This coat belonged to Paramarine Coyle J. O'Neal . (C. Mason)

Right: The parachutist coat in wear. In this photograph, four Paramarines are crossing one of the innumerable creeks and streams on Bougainville in 1943. All are wearing the parachutist utility uniform with the distinctive slanted pockets. (USMC)

The back of the parachutist utility coat. (C. Mason)

The back of the brown side. (C. Mason)

The brown side of the parachutist utility coat. (C. Mason)

Right: Close up view of the neck area showing O'Neal's name stenciled underneath the collar. (C. Mason)

Another parachutist coat. (L. Jacobsen)

The back of the coat. (L. Jacobsen)

A third parachutist utility coat. (J. Busardo)

The back of the coat. (J. Busardo)

Opposite: Paramarines John E. Smith of Company L, 3rd Marine Parachute Battalion (left) and Raymond Barz of Company A, 3rd Marine Parachute Battalion (right) wearing the parachutist utility uniform on New Caledonia. Both are armed with the Johnson light machine gun and both are wearing Marine Corps paratrooper boots. (D. Bailey)

Left: form ("Trousers, utility, HBT, camouflage, [parachutist.]"). Very few original examples of these trousers survive today. The author is aware of just three. (L. Jacobsen)

Below: The back of the parachutist trousers, green side out. (L. Jacobsen)

The front of the parachutist trousers shown brown side out. (L. Jacobsen)

The back of the parachutist pants, shown brown side out. (L. Jacobsen)

Above: A close up of the green side, front, showing the pockets and fly closure.
Below: This close up shows the waistband folded back. (L. Jacobsen)

Above: Paramarines firing at Japanese snipers in the jungle on Koiari Beach. The rear pocket flaps and snaps of the parachutist trousers are clearly visible on the Paramarine on the left. Note the cases of K rations piled up on the beach on the right and the captured Japanese field gun on the far left. (USMC)

Below: A close up of the seat of the trousers, shown green side out. (L. Jacobsen)

Above: The front of the parachutist pants, in detail, brown side out. (L. Jacobsen)
Below: The seat of the parachutist utility trousers, brown side out. (L. Jacobsen)

The first model of the "parachutist coverall," or jump smock. The smock, designed by Captain Robert H. Williams, USMC, echoes the design of the early German paratrooper smocks (as did the British design of

World War II), reflecting the close Allied scrutiny of the German program and equipment. This smock belonged to Paramarine instructor Coyle J. O'Neal. (C. Mason)

Left: The smocks were considered as jump and training equipment, and as Marine Parachute units in the Pacific Theatre never made an operational jump the smocks were not worn in action. This photograph shows three Marines on Tarawa examining destroyed Japanese equipment. The Marine with his back to the camera is clearly wearing a first model smock, probably shortened into a jacket. The explanation of how he got this garment is unknown. (USMC)

Bottom: A close-up photo of the chest of the first model smock. The small rectangular left chest pocket was designed to hold a D Ration. (C. Mason)

Below: A standard issue D Ration. (C. Mason)

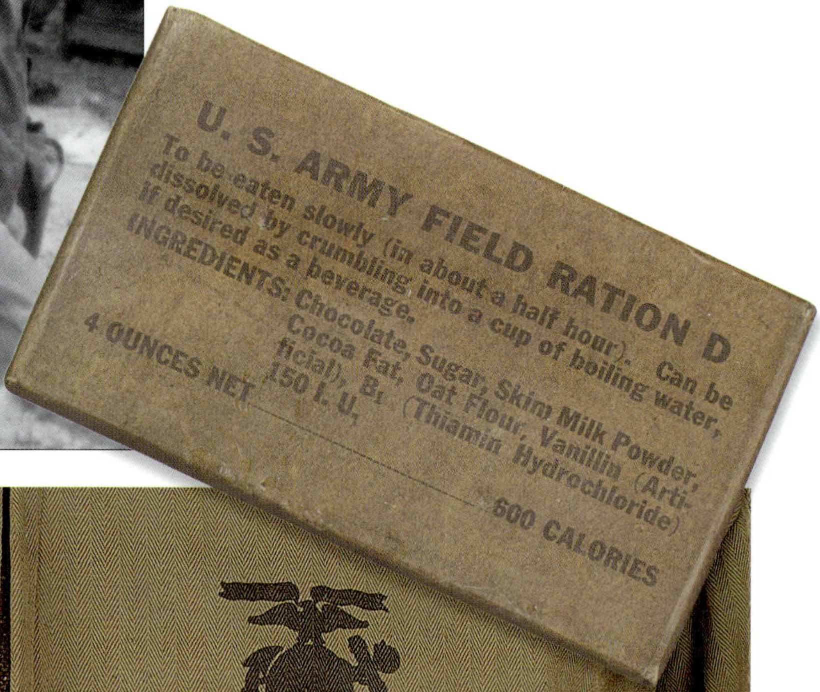

Above: A detail shot of the padded elbow of the first model smock. The Marine design incorporated elbow padding, whereas the Germans favored separate, removable elbow pads for the jumper. (C. Mason)

Right: A close up of the first model smock's cargo and parachutist knife pockets. Very early production smocks lacked these features. (C. Mason)

Above right: Paramarines in training wearing the first model smock. (USMC)

A mint, unissued example of the first model smock. (J. Busardo)

The reverse of the unissued example. (J. Busardo)

USMC

Opposite
Top: A detail of the chest of this smock. (J. Busardo)
Bottom left: A close up of the elbow area. (J. Busardo)
Bottom right: The details of the leg pocket. (J. Busardo)

Above: This Paramarine is handling a block of explosives during training in demolitions. He has sewn Corporal chevrons to his first pattern jump smock. (USMC)

Three photos of an experimental prototype camouflage first model smock, which was never produced. (National Archives)

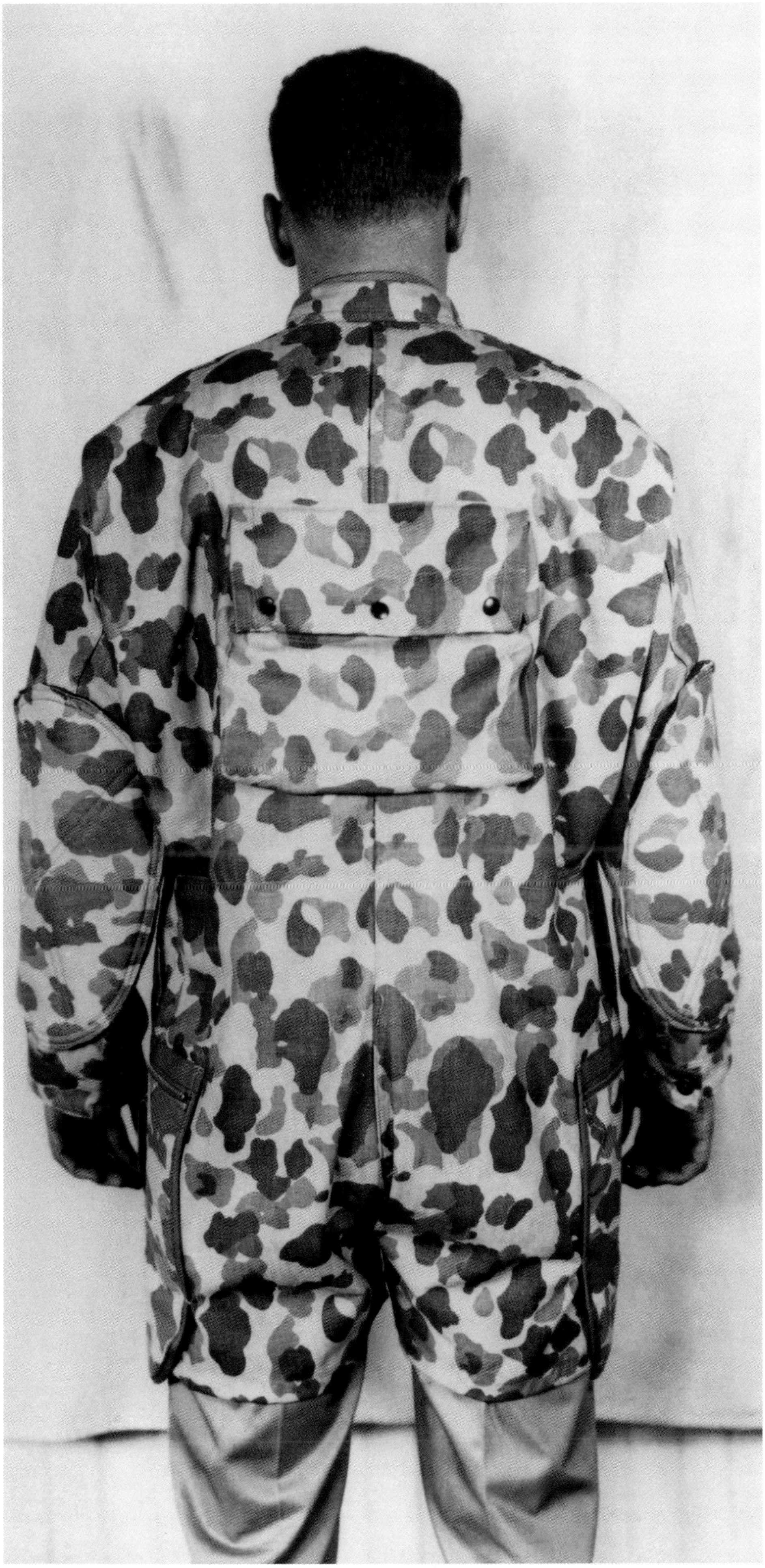

The second model parachutist smock, front view, green side out. The slanted pockets on the second model smock match the slanted pockets of the parachutist's utility coat over which the smock was to be worn when jumping. (J. Busardo)

The back side of the same smock showing the two flapped zipper openings to the large back pocket, believed to be designed to hold a poncho or blanket. (J. Busardo)

The front of the second model smock, brown side out. (J. Busardo)

Opposite
Top: A close up of the chest area of the second model smock showing the slanting chest pockets. (J. Busardo)

Bottom left: A detail view of the snap and zip rear pocket opening in the closed position. The second model smock had integral, concealed elbow pads, as seen here. (J. Busardo)

Bottom right: The hip pocket is shown here in an open position. (J. Busardo)

The back side of the same smock, brown side out. (J. Busardo)

A close up view of the chest area, brown side out. (J. Busardo)

Private First Class Kenneth K. Kleist, Headquarters Company, 3rd Marine Parachute Battalion, wearing the second model smock in training in February 1943.

This painting, entitled "Paramarine," was painted by the renowned American artist Harry Jackson as a Private First Class and war artist in the Marine Corps during World War II. The title of the painting demonstrates how the term "Paramarine" remained in common usage within the Marine Corps well into the war. (USMC and Harry Jackson)

The third model parachutist smock. The third model was much simpler in design than the second model. The two upper and two lower patch pockets are gone, replaced by internal pockets with four snaps, located

along the front closure. This smock belonged to Marine paratrooper Gordon C. Still, Company K, 3rd Marine Parachute Battalion. Still was wounded in action on Iwo Jima with 28th Marines. (J. Busardo)

Below: A detail view of the elbow pad on the third model, now sewn on externally, rather than having the hidden padding of the second model. (J. Busardo)

The third model smock, front view, brown side out. (J. Busardo)

The rear of the third model smock, brown side out. (J. Busardo)

A close-up of the left (brown side) shoulder stenciled "G.C. Still." (J. Busardo)

Below: The simplified access to the rear pocket on the third model smock (brown side), now moved to the side seam. (J. Busardo)

Bottom right: The external padded elbow area of the third model smock (brown side). (J. Busardo)

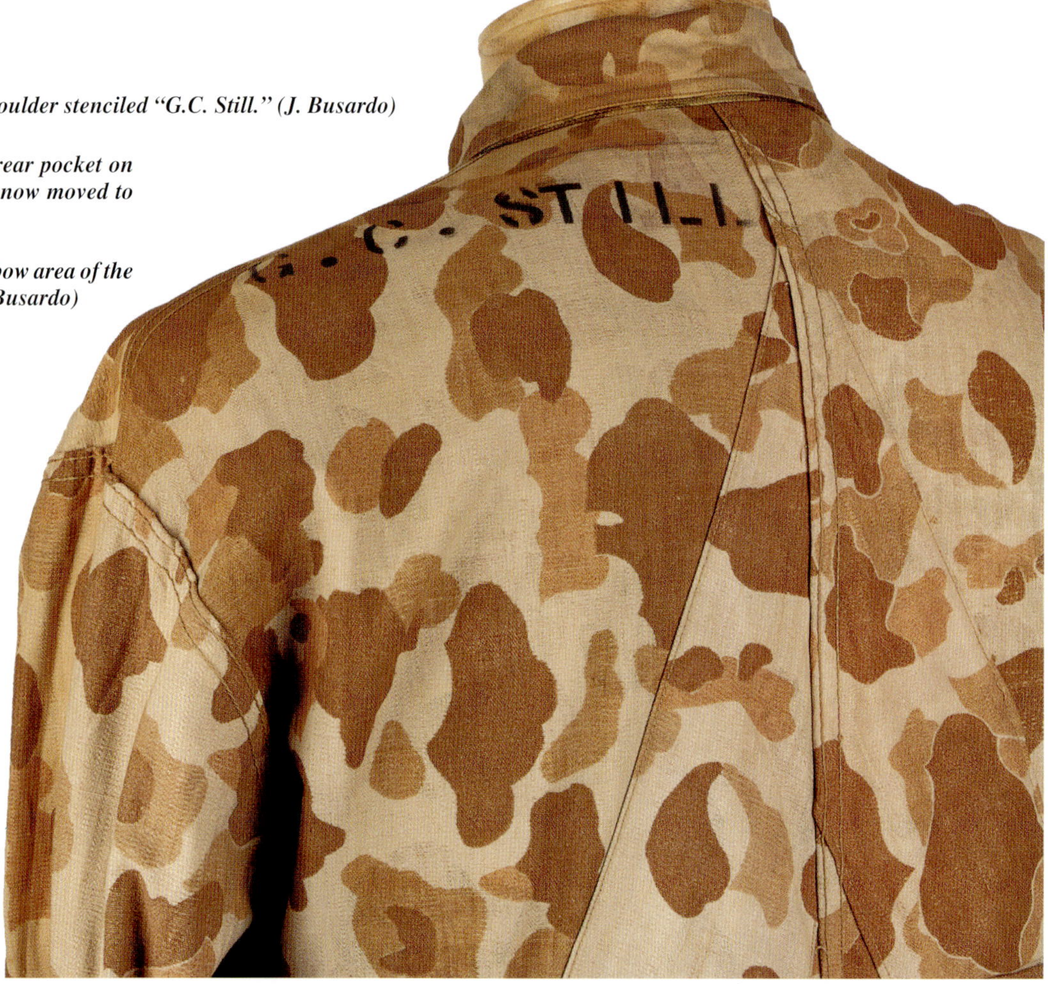

A group of Marines examine Japanese souvenirs after the battle of Tarawa. The Marine on the far left wears an unidentified camouflage garment which appears too long to be a camouflage utility shirt and might be a modified parachutist smock. (USMC)

One piece Marine Corps coveralls made of light green cotton herringbone twill (HBT) material. (L. Jacobsen)

A Paramarine standing in the door of an aircraft during training wearing the one piece coveralls (USMC).

These coveralls were made by sewing an HBT shirt and trousers together later in the war. They belonged to Marine paratrooper Coyle O'Neal while he was serving as an instructor. (C. Mason)

This photograph of Paramarine trainees in San Diego clearly shows the use of the white name tapes, which were worn over the left pocket of the jumpsuit in training at both Lakehurst and Camp Gillespie. (USMC)

A Model 43 (M43) field jacket which belonged to Marine paratrooper Sergeant Harold A. Hoeschen. (L. Jacobsen)

The back of Hoeschen's M43 jacket. Note the stenciled Paramarine insignia. (L. Jacobsen)

A Marine on Iwo Jima wearing an M41 field jacket examines a dud Japanese rocket (USMC).

A leather flight suit shortened to a jacket which belonged to Marine paratrooper Sergeant Harold A Hoeschen while serving as an instructor. (L. Jacobsen)

A view of the back of the jacket, stenciled "NAS MPLS," believed to be the official abbreviation for Naval Air Station, Minneapolis, the location of a Marine Corps school for parachute riggers during the war. (L. Jacobsen)

Hoeschen's initials "HAH," written in ink, are just visible in the lining. (L. Jacobsen)

A detail of the right breast, showing the word "PARA" stenciled in yellow. (L. Jacobsen)

CHAPTER TWO

Headgear

An early M1 helmet of the type worn by Marine parachutists. The Marines seldom used the special airborne liner of the Army with its four-point liner chinstrap attachment design. (C. Mason)

This Marine Corps publicity photo shows a Paramarine in a first model jump smock wearing the Army style airborne liner system. (USMC via G. Petersen)

Two Paramarine mortermen also wearing the Army style airborne liner, rarely observed in Marine Corps use during the war. (USMC via G. Petersen)

An early Hawley liner for the M1 helmet. (C. Mason)

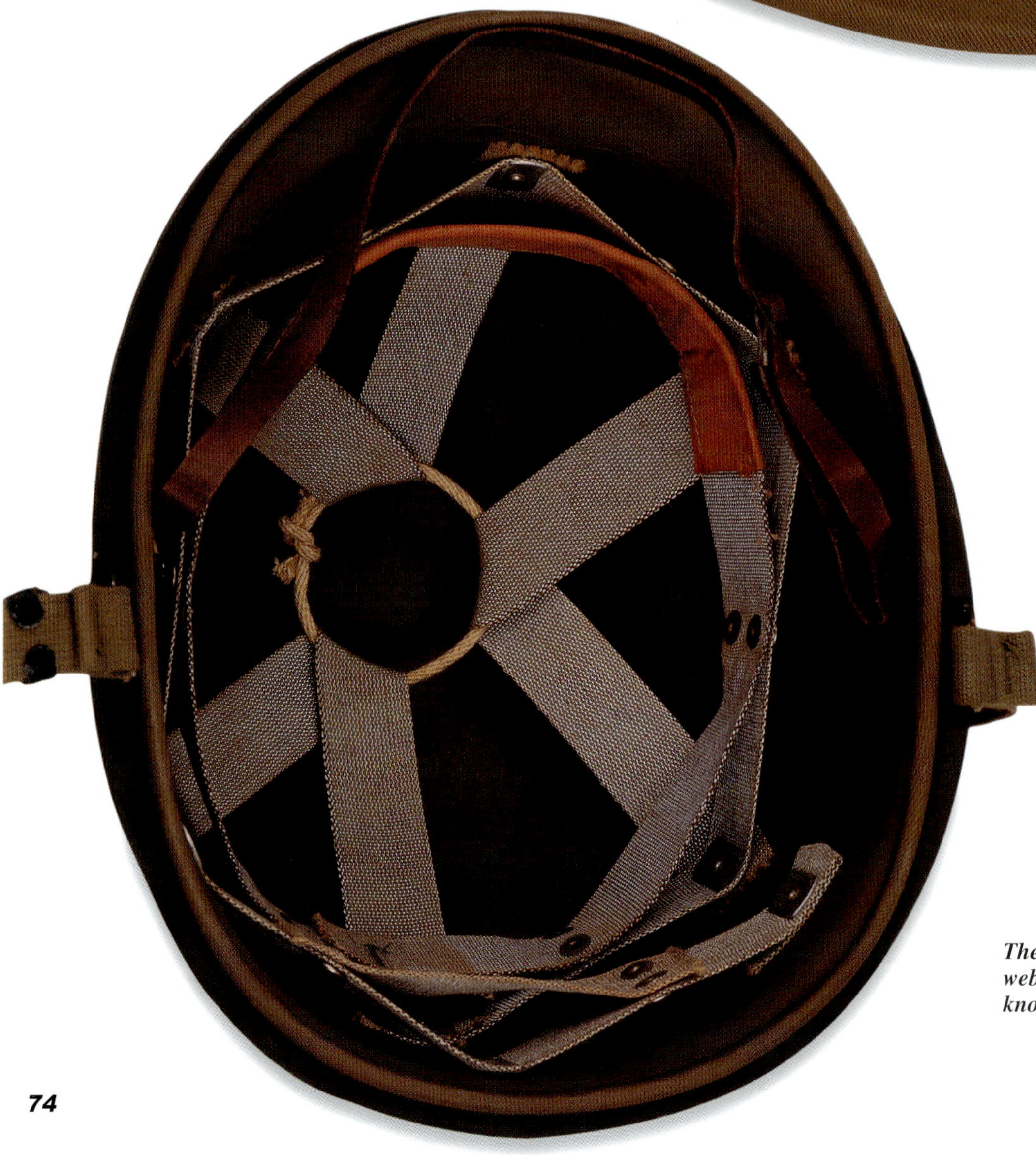

The interior of the early liners had blue webbing, eventually replaced by the well-known khaki webbing color. (C. Mason)

Marine Parachutist Private First Class Gilbert Dionne (a third cousin of the famous Dionne quintuplets of the 1940s) wears the M-1 helmet over a leather flight helmet in 1943. Dionne served in Company G, 2nd Marine Parachute Battalion. He was also wounded on Iwo Jima with 27th Marines. (USMC via Leif Jacobsen.)

Paramarine twins Elba (left) and Delba (right) Lewis wearing the Hawley liner while in training. The Lewis brothers hailed from Franklin, Indiana and served in the same squad of the same platoon of Headquarters Company, 2nd Marine Parachute Battalion. Both wear the Model P1941 HBT Utility Coat. Elba's HBT jacket is stenciled with the EGA and USMC on the pocket, his brother Delba's is not (Perhaps their squad mates told them apart this way). Delba was wounded on Iwo Jima with 27th Marines. (USMC)

A Navy and Marine Corps leather flight helmet design often worn by Marine parachutists in training. The flight helmet was sometimes worn alone while jumping, and sometimes worn under the M1 helmet while jumping. (C. Mason)

A detail of the scarce Marine Corps issue flight helmet, showing the letters "USMC" pressed into the leather in the back. (C. Mason)

Cloth flight helmets similar to this AAF model were also worn. (C. Mason)

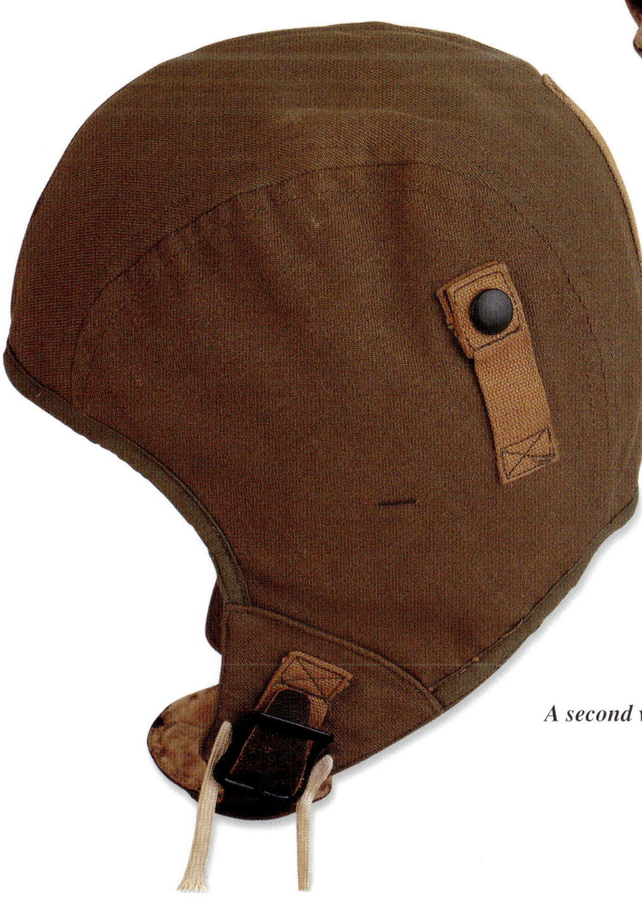

A second view of the cloth flight helmet. (C. Mason)

A group of Paramarines in training at Lakehurst wear coveralls and a mix of leather and cloth flight helmets (USMC).

USMC

107

Each man hooks parachute rip cord to cable. As he falls free from open door his 'chute is thus automatically pulled open.

Men leave plane speedily, 10 jumping within 12 seconds. Note boots of one man as another falls out and a third moves up.

Although men carry emergency chest 'chute, automatic pull (seen below in action) rarely fails to operate properly.

This sequence of photographs shows Paramarines jumping in training while wearing only cloth and leather flight helmets. (USMC)

Camp Elliott, 1943. Second Lieutenant James P. Gass wears the leather flight helmet under the standard M-1 steel helmet. He is wearing the second model smock and the two piece camouflage parachutist utility uniform and is armed with the M55 Reising submachine gun with folding stock. Gass served with Headquarters Company, 3rd Marine Parachute Battalion and later with 27th Marines on Iwo Jima, where he earned the Bronze Star. (USMC)

Right: The camouflage helmet cover, first model, without slits for foliage. Note that there is no EGA emblem stenciled on the front. Most USMC collectors agree that helmet covers were not stenciled during WWII, and that the stenciling of the emblem was added to existing stocks of WWII era-covers during the Korean War. (C. Mason)

Below: This photograph of Corporal Robert Foster on Tarawa affords an excellent view of the second model helmet cover, clearly without an EGA, and the early Hawley liner worn underneath. (USMC)

An angle view of the same cover. (C. Mason)

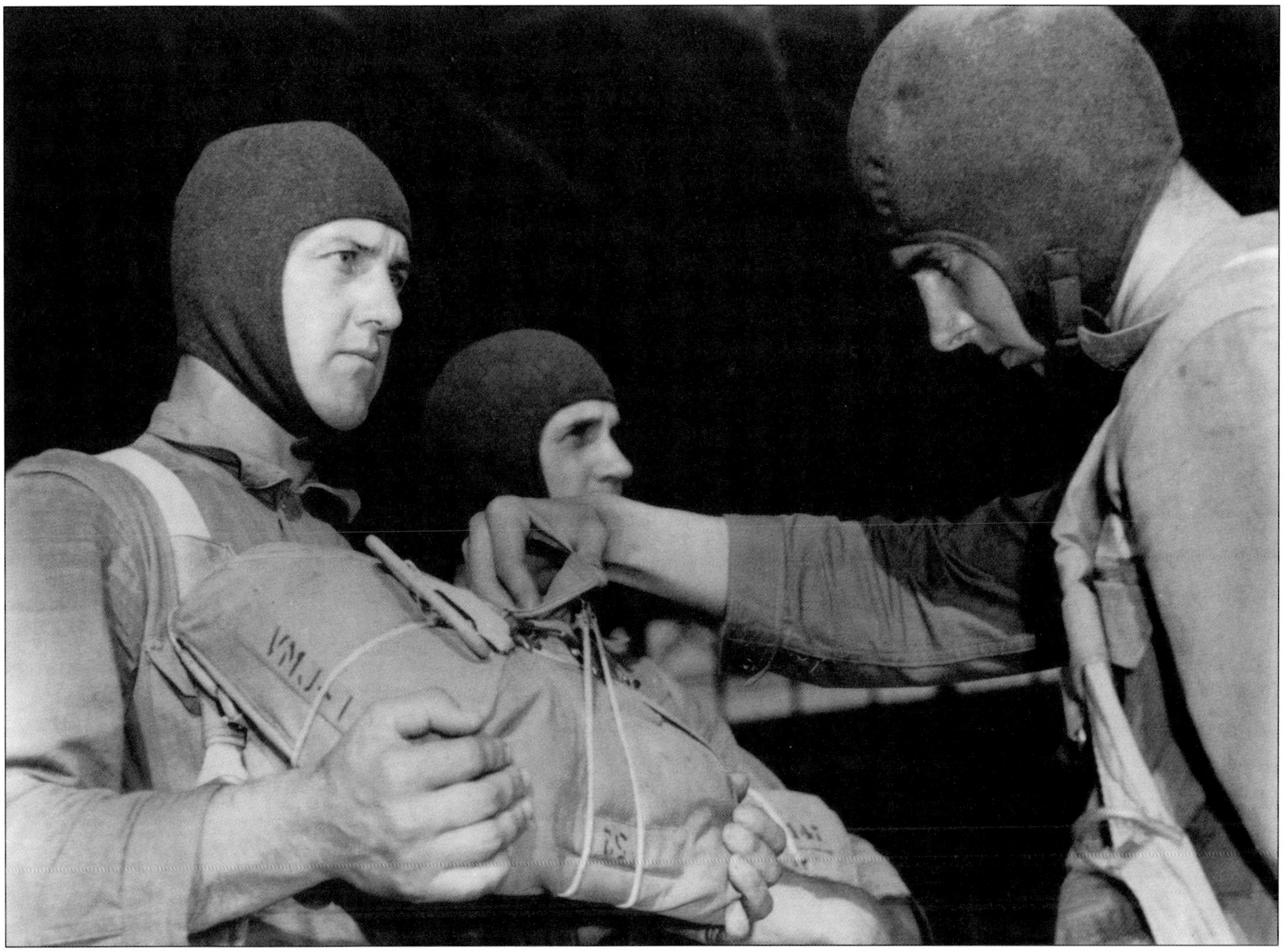

Above: Paramarines in training at Lakehurst wear cloth flight helmets. (USMC)
Below: Marines in a war dog platoon on Iwo Jima wearing first and second model helmet covers. (USMC)

A group of men from Second Marines on Saipan on the move. The man closest to the camera, among others, wears the helmet cover with the rear flaps down. (USMC)

Above: A group of Marine mortarmen on Saipan wearing the helmet cover. The man in the center wears the cover with the rear cloth tongues down as a neck shield instead of tucking them between the helmet and the liner. (USMC)

Left: The markings on one of the controversial "Australian Made" USMC helmet covers. Although the cover appears in every respect to be of the period, some collectors disagree over their authenticity on the grounds that no records can be found of this company making covers. (J. Busardo)

Opposite A group of Marines and Corpsmen at an aid station on Okinawa. The Marine in the middle ground wears a rarely seen camouflage net over his helmet cover. (USMC)

The khaki summer garrison, or "overseas," cap with an enlisted EGA. (C. Mason)

Above: The forest green winter service garrison cap which belonged to Paramarine Sergeant Harold Hoeschen. Right: Hoeschen's initials "HAH" inside his garrison cap, which was apparently previously worn by Paramarine Dixon A. Yaste, Jr (4th Marine Parachute Battalion). Yaste's initials "DAY" also appear on the sweatband. (L. Jacobsen)

Three Paramarines from the town of Long Beach, California, wearing the khaki garrison cap. (USMC)

Lieutenant Colonel (later Colonel) Charles E. Shepard, Jr, USMC, wears the forest green winter garrison cap. (C. Mason)

Below: An EGA made by the firm of K C Luke in Australia on a garrison cap. K C Luke made EGA's are among the rarest and most sought after USMC insignia of the war. (C. Mason)

Below right: The reverse of the K C Luke insignia, showing the pin design unique to this firm. (C. Mason)

Miniature jump wings on the right hand side of the garrison cap. This practice was strictly against regulations, but period photos show it was done occasionally by individual Paramarines. This cap was worn by Paramarine Private First Class Edward L. McClure. Below: The Australian maker's label inside McClure's

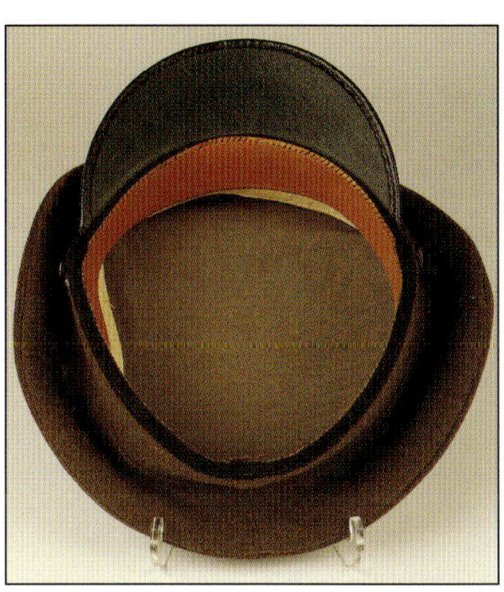

An example of the World War II enlisted barracks cover ("Cap, frame, service, winter, enlisted men's") in forest green with a saddle shape. Left: The interior of this cover. (J. Busardo)

An example of the World War II enlisted barracks cover in the more common standard form. Left: The interior of the standard form cover. (C. Mason)

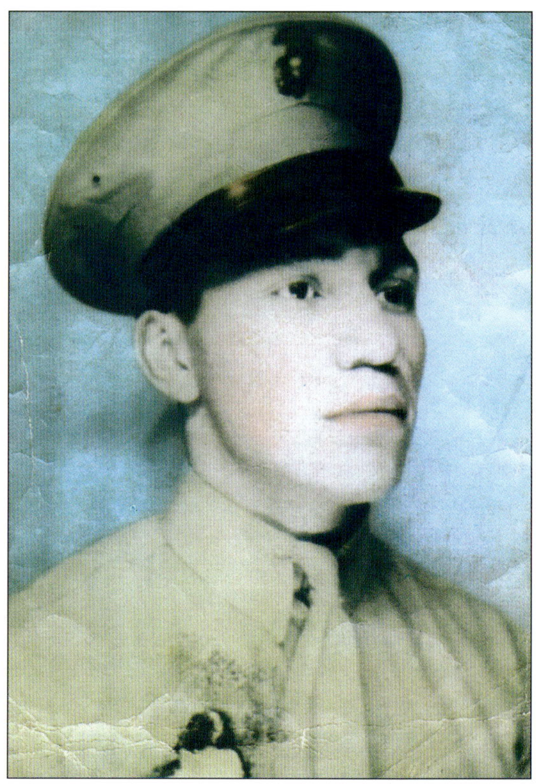

Above: An example of the P1941 sage green billed utility cap worn by a Marine Captain during World War II. It was made from cotton HBT material. (C. Mason)

Right: Demining operations on Iwo Jima. The Marine at left wears a metal EGA insignia on his P1941 short-billed utility cap. The Marine in the center wears a P1944 utility cap with the characteristic longer bill and stenciled EGA. (USMC)

Opposite
Bottom left: An unidentified Paramarine wears the enlisted barracks cover with his forest green winter service uniform. This photograph is proof that while Army paratroopers thought they got the prettiest girls, Paramarines actually did so. (C. Mason)

Bottom right: Paramarine and Seneca Indian Ken Parker from the Cattarlaugus Indian Reservation near Buffalo, New York, wears the barracks cover with summer khaki cover. Parker served in Company C of the 1st Marine Parachute Battalion. He was killed in action on Iwo Jima with the 27th Marines. (L. Jacobsen)

Above: Marine First Lieutenant R. A. Montgomery of Mississippi checks the body of a dead Japanese on Okinawa in April 1945. He is wearing the P1941 utility cap (USMC).

Below: Marines on Saipan operate a captured Japanese field gun against its former owners. Two of them are wearing billed utility caps clearly made from camouflage material. (USMC)

A Marine comforts a buddy who has just seen his best friend killed in action. The kneeling Marine wears a metal EGA on the bill of his mechanics cap rather than the crown. (USMC)

CHAPTER THREE

Equipment

Paramarine standing in a door mock-up during a portion of ground training. Note the details of the TB Dual Purpose Parachute, the jump boots, and the early Hawley helmet liner. (USMC)

The T-5 Main Parachute shown front view. Former Paramarine officers Colonel Dave Severance and Lieutenant Colonel Frank Clagett, who served as the Paraloft Officer at both Lakehurst and Camp Gillespie, report that the Marine Corps used a parachute designated the "TB Dual Purpose Parachute" in training. "TB" stood for "Training, Back Type." It could be opened either by ripcord or static line. (J. Busardo)

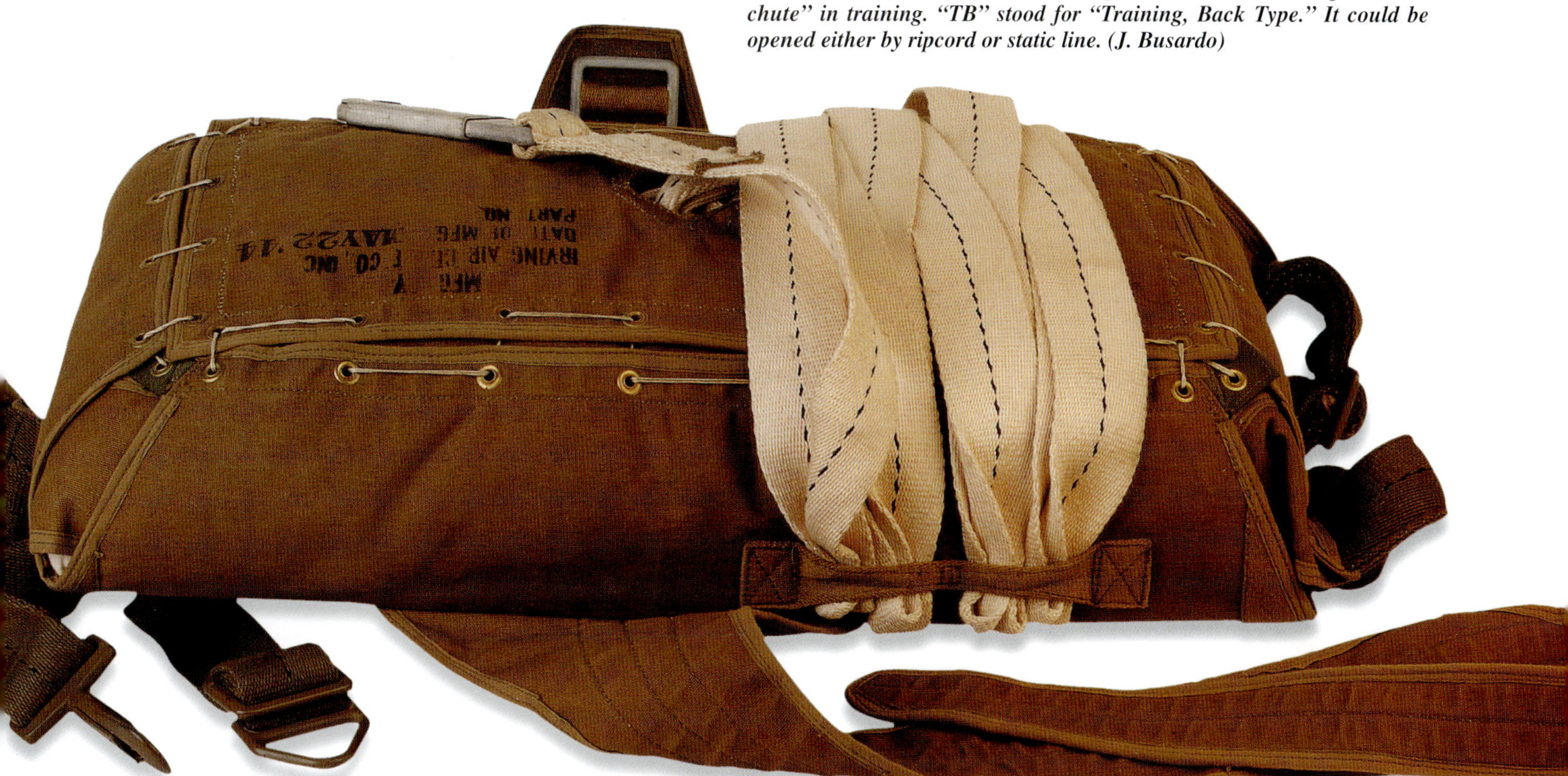

T-5 Main Parachute, rear view. (J. Busardo)

T-5 Main Parachute, side view. (J. Busardo)

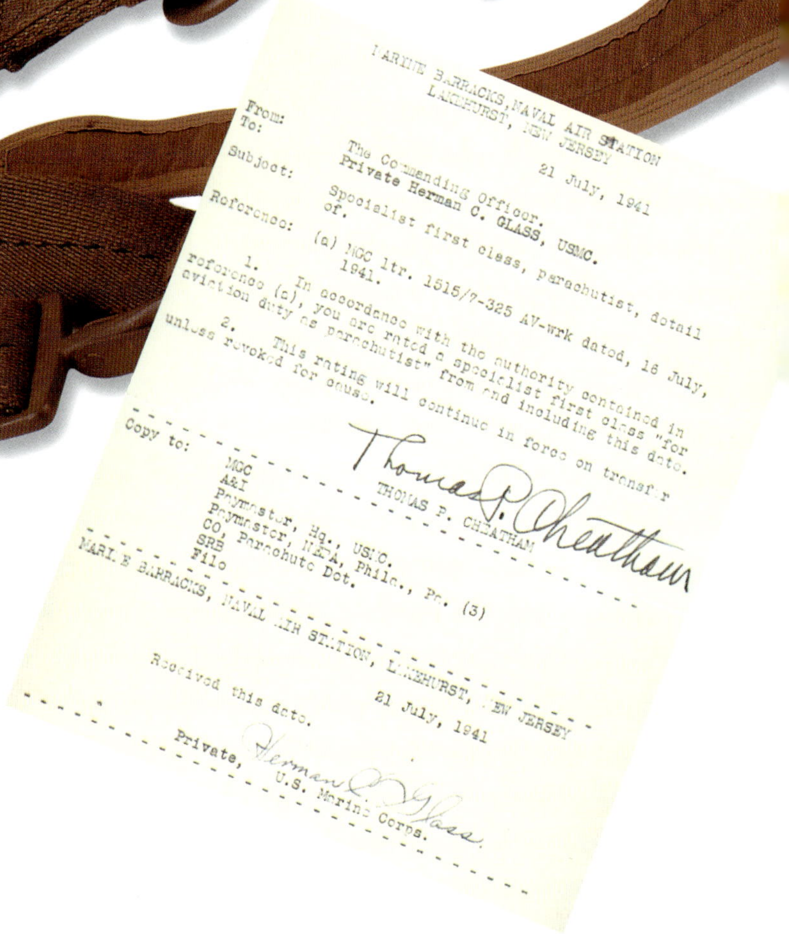

MARINE BARRACKS, NAVAL AIR STATION
LAKEHURST, NEW JERSEY

21 July, 1941

From: The Commanding Officer.
To: Private Herman C. GLASS, USMC.
Subject: Specialist first class, parachutist, detail of.
Reference: (a) MGC ltr. 1515/7-325 AV-wrk dated, 16 July, 1941.

1. In accordance with the authority contained in reference (a), you are rated a specialist first class "for aviation duty as parachutist" from and including this date.
2. This rating will continue in force on transfer unless revoked for cause.

THOMAS P. CHEATHAM

Copy to:
MGC
A&I
Paymaster, Hq., USMC.
Paymaster, MDPA, Phila., Pa. (3)
CO, Parachute Det.
SRB
File

MARINE BARRACKS, NAVAL AIR STATION, LAKEHURST, NEW JERSEY

21 July, 1941

Received this date.

Private, U.S. Marine Corps.

Right: A personnel document from early in the program assigning the rate of "Specialist First Class, parachutist" to Private Hermann C. Glass. Pearl Harbor was still four and a half months in the future when this document was signed. (E. Queen)

The T-5 reserve parachute, front view. The Marine Corps designation for this parachute was "TC, " for "Training, Chest." (J. Busardo)

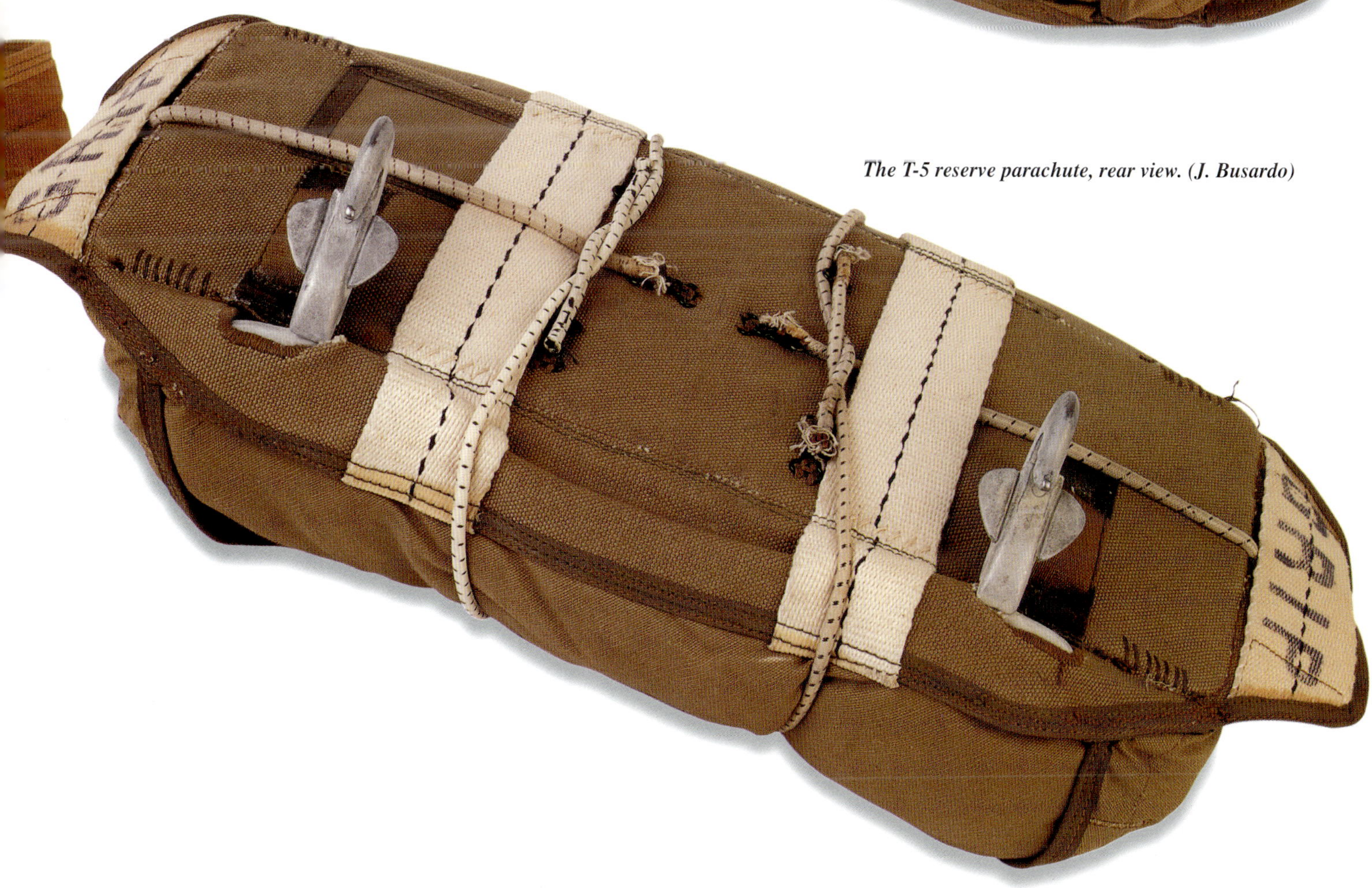

The T-5 reserve parachute, rear view. (J. Busardo)

Left: This photo clearly shows the TB Dual Purpose Parachute parachute capable of both ripcord and static line deployment. The ripcord guide is in position over his shoulder, and the static line is clearly visible going back to the fore and aft attachment cable in the fuselage of the aircraft. (USMC)

Inset right: Paramarines on a tactical training jump using the TB Dual Purpose Parachute, clearly identifiable by the presence of the little pilot parachutes above the main canopies over several of the descending men. The small pilot parachute served to help pull the main canopy out of the deployment bag when it was ripcord activated, but served no purpose when the static line pulled the canopy out of the deployment bag. (L. Jacobsen)

Background: Paramarines wearing the TB Dual Purpose Parachute run to board an aircraft for a training jump. (USMC)

This World War II advertisement for Camel cigarettes appeared in many magazines during the period. It shows a Marine parachutist wearing the camouflage jump smock, parachute and Reising submachine gun "chuted-up" for a jump.

Paramarines on a work detail load parachutes for transport. (USMC)

A group of Marine parachutist trainees is checked by a jumpmaster before boarding their aircraft for a training jump. (USMC)

A group of Paramarines. From left to right, Lieutenant Blasingame, Gunnery Sergeant Pettigrew, and Platoon Sergeant Viladoga with their mascot before a training jump. They are wearing TB Dual Purpose parachutes. (USMC)

Left: Toughening up: A group of Marine parachutist trainees carry the rolled up gymnastics mat they will practice their Parachute Landing Falls (PLFs) on as part of the their training in landing safely. Right: Paramarine trainees practice PLFs. (USMC)

Left: Another photo of trainees practicing their PLFs. This was repeated hundreds of times, until it became second nature. Right: Trainees practicing their door exits. (USMC)

Left: A "Chalk Talk" for a group of Paramarine trainees, a welcome break from the rigorous and almost continuous physical activities of training. Right: Parachute training. The parachute harness and suspension are clearly seen in this photograph. (USMC)

This commercial parachute tower at Hightstown, New Jersey, was originally part of the New York World's Fair in 1939. It was subsequently purchased and operated as an amusement park ride for civilians by the Safe Parachute Jump Company. It was then acquired by the military for parachute training. Here a test parachutist tests the tower before the purchase for suitability as a training device. As an amusement ride, it had a seat for the passenger and had to be adapted to a military parachute harness. (USMC)

Right: A Marine test parachutist at Hightstown after landing wearing a dummy parachute pack. The parachute was held open by a wire frame and released from the top of the tower, so the backpack was for testing purposes only. The Marine has kept his overseas cap on for the jump. (USMC)

Below: This test parachutist trying out the tower prior to acquisition is clearly an enlisted Sailor from his dungaree uniform, complete with correct issue white socks and low-quarter black street shoes. (USMC)

Training in the method for collapsing the canopy in the event of being dragged across a drop zone by high winds after landing. (USMC)

Below: After a good landing: A Paramarine trainee field stows his parachute on the drop zone for transport off the drop zone to the paraloft for repacking. (USMC)

A stick of Paramarines wearing the the TB Dual Purpose Parachute studies the drop zone before a practice jump. The static lines are clearly visible over their shoulders. (USMC)

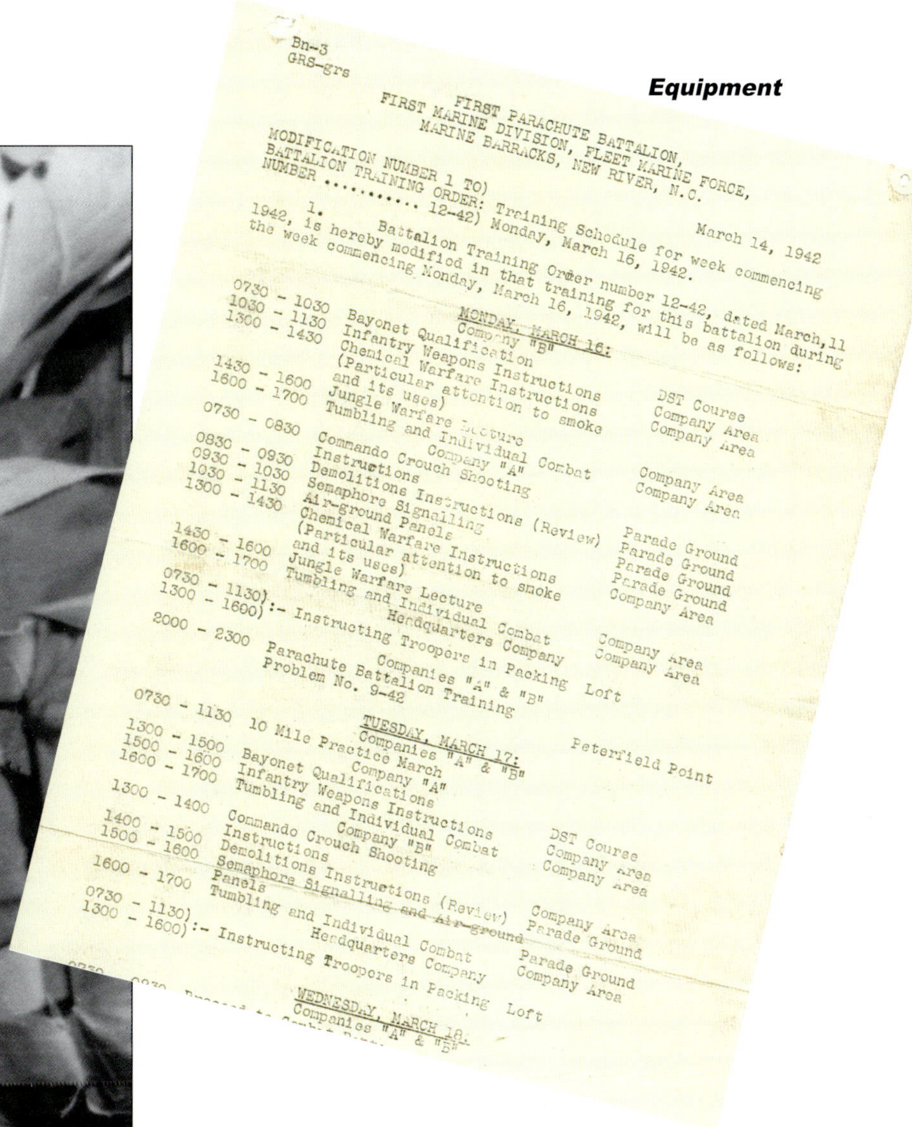

Bn-3
GRS-grs

FIRST PARACHUTE BATTALION,
FIRST MARINE DIVISION, FLEET MARINE FORCE,
MARINE BARRACKS, NEW RIVER, N.C.

March 14, 1942

MODIFICATION NUMBER 1 TO)
BATTALION TRAINING ORDER: Training Schedule for week commencing
NUMBER .... 12-42) Monday, March 16, 1942.

1. Battalion Training Order number 12-42, dated March, 11 1942, is hereby modified in that training for this battalion during the week commencing Monday, March 16, 1942, will be as follows:

MONDAY, MARCH 16:
Company "B"
0730 - 1030 Bayonet Qualification — DST Course
1030 - 1130 Infantry Weapons Instructions — Company Area
1300 - 1430 Chemical Warfare Instructions (Particular attention to smoke and its uses) — Company Area
1430 - 1600 Jungle Warfare Lecture
1600 - 1700 Tumbling and Individual Combat — Company Area
Company "A"
0730 - 0830 Commando Crouch Shooting — Company Area
0830 - 0930 Instructions
0930 - 1030 Demolitions
1030 - 1130 Semaphore Instructions (Review) — Parade Ground
1300 - 1430 Semaphore Signalling — Parade Ground
Air-ground Panels — Parade Ground
Chemical Warfare Instructions (Particular attention to smoke and its uses) — Parade Ground
1430 - 1600 Jungle Warfare Lecture — Company Area
1600 - 1700 Tumbling and Individual Combat
Headquarters Company
0730 - 1130)
1300 - 1600):- Instructing Troopers in Packing Loft — Company Area
Companies "A" & "B"
2000 - 2300 Parachute Battalion Training Problem No. 9-42 — Company Area, Peterfield Point

TUESDAY, MARCH 17:
Companies "A" & "B"
0730 - 1130 10 Mile Practice March
Company "A"
1300 - 1500 Bayonet Qualifications
1500 - 1600 Infantry Weapons Instructions — DST Course
1600 - 1700 Tumbling and Individual Combat — Company Area
Company "B"
1300 - 1400 Commando Crouch Shooting — Company Area
1400 - 1500 Instructions
1500 - 1600 Demolitions Instructions (Review) — Company Area
Semaphore Signalling and Air-ground Panels — Parade Ground
1600 - 1700 Tumbling and Individual Combat — Parade Ground
Headquarters Company
0730 - 1130)
1300 - 1600):- Instructing Troopers in Packing Loft — Company Area

WEDNESDAY, MARCH 18:
Companies "A" & "B"

This training schedule for the week of March 16-20, 1942, for the 1st Marine Parachute Battalion, gives a sense of the intensity of Paramarine training. It was brought home by Marine parachutist Edward Blenski, Company B, 1st Marine Parachute Battalion. Blenski fought on Gavutu. (L. Jacobsen)

Left: Paramarine instructor Coyle J. O'Neal after a tree landing with the T-5 parachute. This kind of landing, over which the World War II-era jumper had little control, was generally a result of where the jumpmaster in the aircraft "spotted" the stick, as World War II parachutes were even less maneuverable than today's modified T-10 troop parachutes. (C. Mason)

Above: Another bad landing. Here Paramarine schools instructor Coyle O'Neal gets his feet wet. (C. Mason)

Marine parachutist jump boots first issued in 1941. Note the rough side of the leather is facing out, as it does on USMC "boondockers," (hence the term "rough-side-out" or simply "rough-out') and the absence of the toe caps found on Army jump boots on these early examples. These boots belonged to Paramarine Erwin Albers, Weapons Company, 1st Marine Parachute Regiment. Albers earned the Bronze Star on Iwo Jima with the 5th Service Battalion. (L. Jacobsen)

Marines also made some use during the war of the Army design with toe caps. (L. Jacobsen)

Above left: Paramarine Corporal Charles C. Clow, Jr, poses for a photograph with a hometown friend in the Navy while on leave in Minneapolis. His winter service uniform trousers have been bloused "paratrooper style" over his spit-shined army style paratrooper boots with toe caps. (L. Jacobsen)

Left: A Paramarine on a rope swing wearing jump boots. (USMC)

Above: A Paramarine training at New River, North Carolina wearing "rough out" Paramarine jump boots. (USMC)

Opposite
Top: Wounded being evacuated off Iwo Jima. Most former members of the 1st Marine Parachute Regiment were transferred to the 5th Marine Division in March 1944 after the parachute force was disbanded and fought on Iwo Jima. Members of 1st Battalion went into the newly-forming 26th Marines, members of the 2nd Battalion went into the new 27th Marines, and members of the 3rd Battalion formed part of the nascent 28th Marines, which together with supporting units formed the Fifth Marine Division. The man sitting farthest left in this photograph is apparently a former Paramarine who has retained his jump boots. (USMC).

Bottom: Apparently the boots somehow found their way to other Marines, too, like the first model smock seen on Tarawa in the uniform chapter. This Marine on Tarawa is wearing jump boots. (USMC)

Left: An early USMC issue canteen and carrier with standard straps and lift-the-dot fasteners. (J. Busardo)

Below: Another early USMC issue canteen and carrier with standard straps and snaps, marked on the bottom with a Marine's name. Marine Corps canteen carriers had no felt linings and have the USMC belt hook. (J. Busardo)

Above left: The first wartime USMC issue canteen carrier was made with cross straps but was designed without a bottom drainage hole. (J. Busardo)

Above: This was corrected in a subsequent redesign of the canteen carrier with cross straps to include a bottom drainage hole. (J. Busardo)

Left: The black enameled canteen and cup often issued to Marines. (J. Busardo)

Above: Paramarines in training at Camp Gillespie wearing World War I era canteens. This black and white photograph has been hand-colored during the war to produce the full-color effect (note that the white shroud lines have been colored in with the same red the artist applied to the whole parachute.) The parachute colors are fanciful. (USMC via G. Petersen) Below: The Marine at right carries one of each style canteen carrier together with other gear. (USMC)

Marines during the fight for Roi Island. The Marine closest to the camera is wearing a captured Japanese canteen. (USMC).

Above: A Marine writing home after the fighting on Kwajelein carries his first aid pouch on his left hip. (USMC) Overlay: The M1912 USMC issue First Aid Pouch with field dressing. (C. Mason)

The rare Paramarine padded camouflage cover for the entrenching tool. Although unmarked, the design and manufacture of this cover appears to be of industrial level, not rigger, manufacture. (L. Jacobsen)

The reverse of the camouflage cover, showing the securing strap and snap. (L. Jacobsen)

A detail view of the fastening strap and snap. (L. Jacobsen)

A close up view of the reverse, showing the snaps for the fastening strap and the buttoning belt loop. (L. Jacobsen)

A Marine wearing the T-handle shovel gives a drink of water to a wounded buddy on Pelilieu. He carries a rolled-up poncho on top of his musette bag carried in the backpack configuration. (USMC)

Left: USMC issue field bag. This musette bag is a variation design of the M1936 canvas field ("musette") bag without provision for wear as a backpack. Below: A detail showing the shoulder strap. Below left: A close-up view showing the flap and markings. (J. Busardo)

A USMC-issue M1936 musette bag. Clearly visible on the bottom is the snapped tab which permitted the bag to be worn as a backpack by running the single strap through the tab. Otherwise it could also be carried over the shoulder. (J. Busardo)

Close-up of the musette bag showing the shoulder strap details. (J. Busardo)

Detail of the manufacturer's markings on this bag. (J. Busardo)

A darker shade of fabric was used on this musette bag. (L. Jacobsen)

A detail of the markings on this musette bag, manufactured by Boyt. (L. Jacobsen)

The front of this bag with the flap open. (L. Jacobsen)

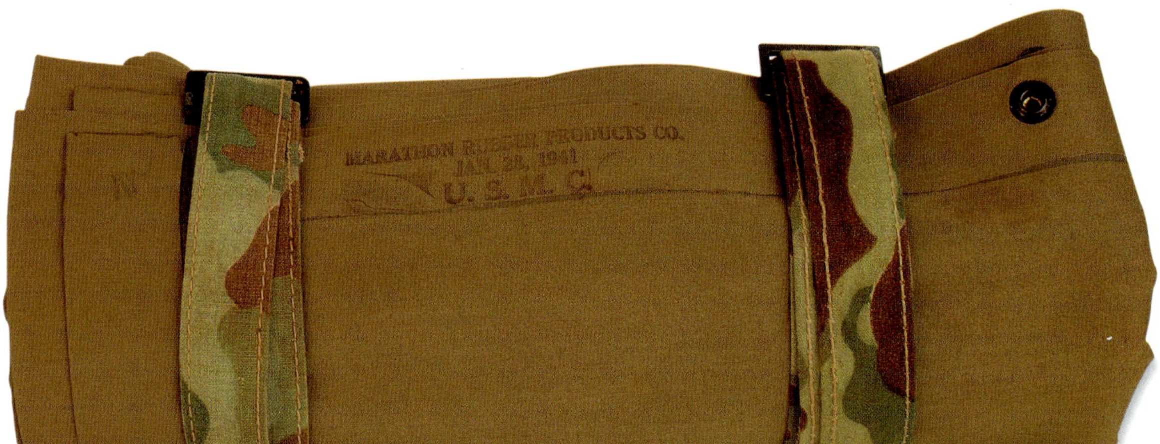

Early USMC poncho rolled and secured by utility straps. This poncho was manufactured by the Marathon Rubber Products Corporation on January 28, 1941. (C. Mason)

A USMC issue pistol belt. (J. Busardo)

A USMC issue M1923 rifle cartridge belt. Originally designed for the 1903 Springfield rifle, it could also carry M-1 Garand rifle clips of eight rounds and was used until after the Korean War. (J. Busardo)

The front of the Army camouflage jungle pack. While on Vella Lavela island, Major Harry Torgerson arranged to trade several cases of alcohol to an Army supply facility for enough jungle material, including packs, boots and hammocks, to equip the entire parachute regiment. (J. Busardo)

The reverse of the jungle pack. (J. Busardo)

Front view of the USMC-issue document (map) case. (J. Busardo)

The reverse of the USMC issue map case. (J. Busardo)

Inside view of a USMC issue map case showing the celluloid map pocket and the felt writing surface which also protected the celluloid, which was easily scratched. (J. Busardo)

A rigger-made pouch carried by a Paramarine. (L. Jacobsen)

Left: A Paramarine Rigger's tool kit. This kit belonged to Paramarine Sergeant Harold Hoeschen, and is stenciled "PARA USMC." (L. Jacobsen)

Below: The interior of the Rigger's Kit, showing pockets for various parachute packing tools. (L. Jacobsen)

Above: The MAB-1 radio designed specifically for Paramarines, showing the front of the khaki case. Below: The reverse of the MAB-1 radio case. (J. Busardo)

The MAB-1 radio components, showing the cloth headset worn under the M-1 helmet. (J. Busardo)

Close-up view showing the inner ear piece. (J. Busardo)

The headset of the MAB-1 radio. (J. Busardo)

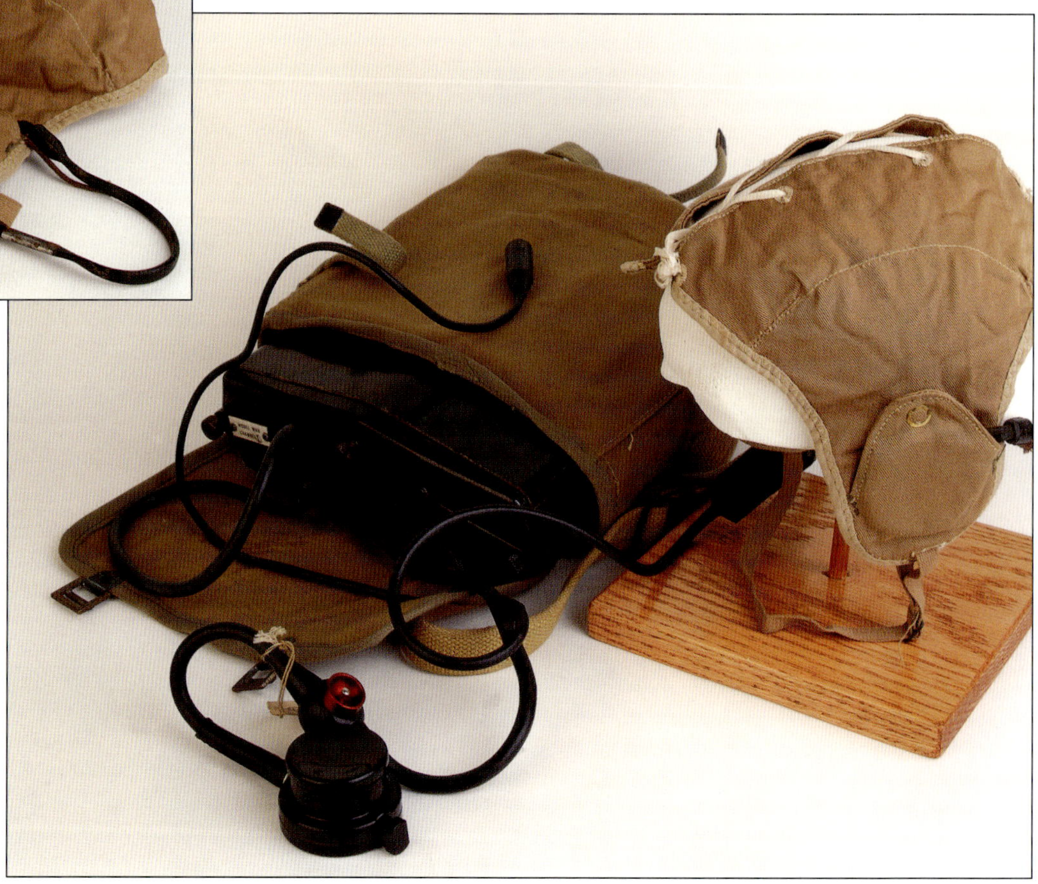

The antenna for the MAB-1 radio. (J. Busardo)

The MAB-1 radio worn in the field by a Marine parachutist. (USMC via L. Jacobsen)

The MAB-1 radio ready to jump, secured behind the reserve parachute. (USMC via L. Jacobsen)

The MAB-1 radio in use, showing the antennae fully extended, and in full rear view. (USMC via L. Jacobsen)

RESTRICTED

INSTRUCTION BOOK
For
NAVY MODEL MAB
RADIO TELEPHONE TRANSMITTING AND
RECEIVING EQUIPMENT

Manufactured For
U. S. NAVY DEPARTMENT
BUREAU OF SHIPS

By
COMMUNICATIONS COMPANY, INC.
Coral Gables, Florida

CONTRACT NXss-14439
Dated: October 1, 1942

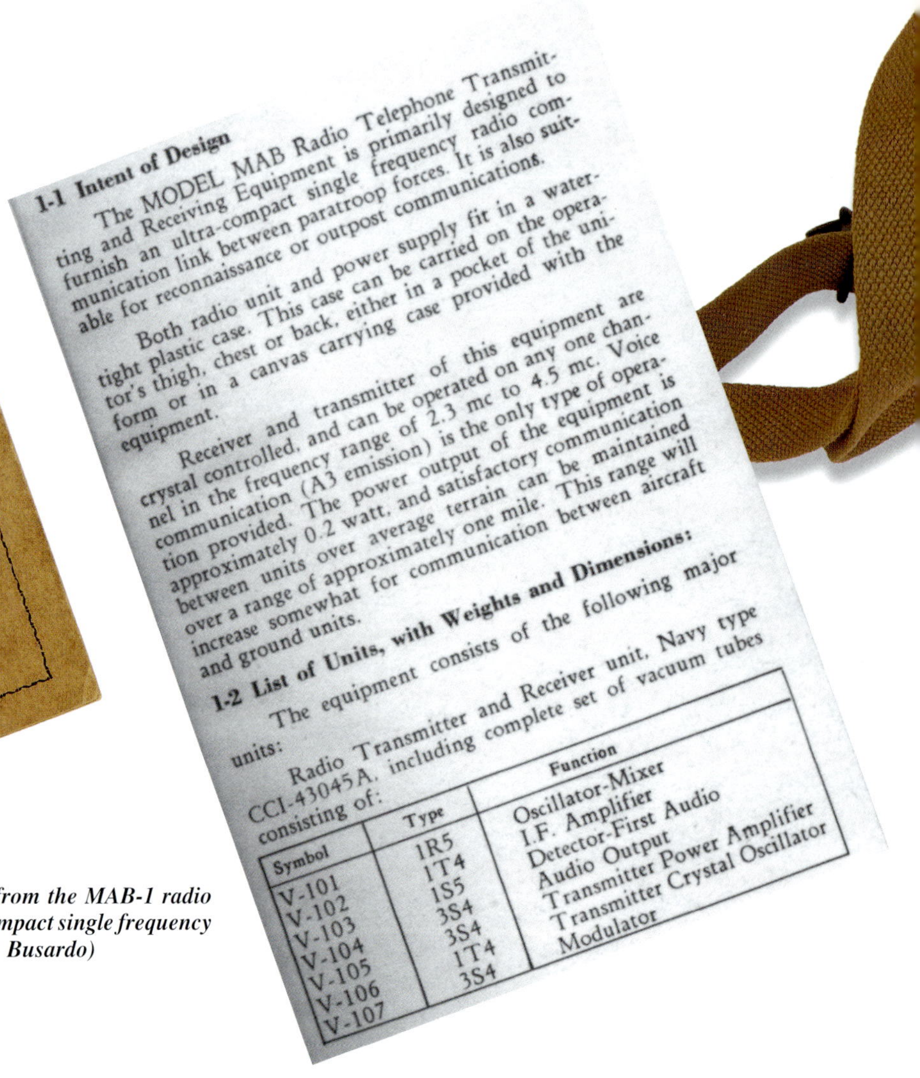

1-1 Intent of Design

The MODEL MAB Radio Telephone Transmitting and Receiving Equipment is primarily designed to furnish an ultra-compact single frequency radio communication link between paratroop forces. It is also suitable for reconnaissance or outpost communications.

Both radio unit and power supply fit in a watertight plastic case. This case can be carried on the operator's thigh, chest or back, either in a pocket of the uniform or in a canvas carrying case provided with the equipment.

Receiver and transmitter of this equipment are crystal controlled, and can be operated on any one channel in the frequency range of 2.3 mc to 4.5 mc. Voice communication (A3 emission) is the only type of operation provided. The power output of the equipment is approximately 0.2 watt, and satisfactory communication between units over average terrain can be maintained over a range of approximately one mile. This range will increase somewhat for communication between aircraft and ground units.

1-2 List of Units, with Weights and Dimensions:

The equipment consists of the following major units:

Radio Transmitter and Receiver unit, Navy type CCI-43045A, including complete set of vacuum tubes consisting of:

Symbol	Type	Function
V-101	1R5	Oscillator-Mixer
V-102	1T4	I.F. Amplifier
V-103	1S5	Detector-First Audio
V-104	3S4	Audio Output
V-105	3S4	Transmitter Power Amplifier
V-106	1T4	Transmitter Crystal Oscillator
V-107	3S4	Modulator

Left: The manual for the MAB-1 radio. Right: A page from the MAB-1 radio manual indicating the "Purpose of the radio is an ultra compact single frequency radio communication link between paratroop forces." (J. Busardo)

Left: The Navy portable CEX-10172 radio issued to Navy Beachmaster and Joint Assault Companies (JASCOs). The radio likely saw use with Marine parachute forces as well. (J. Busardo)

Above: The CEX-10172 radio. (J. Busardo) Below: Detail of the CEX-10172 radio. (J. Busardo)

The Marine Corps parachute supply canister. Loaded with supplies, the canister could be air-dropped to parachute forces in enemy territory. The canister resembles the drop container designed by the German parachute branch before the war. (C. Mason)

The supply canister with wheels attached. The wheels were stowed inside when the canister was dropped, together with a flexible belly-band which held the axles. (C. Mason)

The canister would also stand upright. (C. Mason)

A close-up view of the wheels and axle assembly. (C. Mason)

A detail view of the belly band, which held the axles and wheels firmly in place for transport. (C. Mason)

A close-up view of the removable lid. (C. Mason)

The padded end of the canister and the canvas carrying or towing handle. (C. Mason)

The canister was made of lightweight fiberglass material, and was fitted with a nose cone for aerodynamic purposes. The cone also functioned as a shock absorber on landing. (C. Mason)

CHAPTER FOUR

Weapons

An example of the Marine utility knife issued to all Marines in a combat zone beginning in 1943. The pattern had four makers, but all came to be generically called "Kabars" (pronounced "KAY-bar"), for one of them, the "Ka-Bar Knife" manufactured by the Union Cutlery Company. The particular Marine utility knife shown above was trademarked as the "PAL Knife," and was manufactured by the Pal Blade Company. Camillus Cutlery and Robeson Cutlery (maker of the "Robeson Shuredge") were the other two makers. (C. Mason)

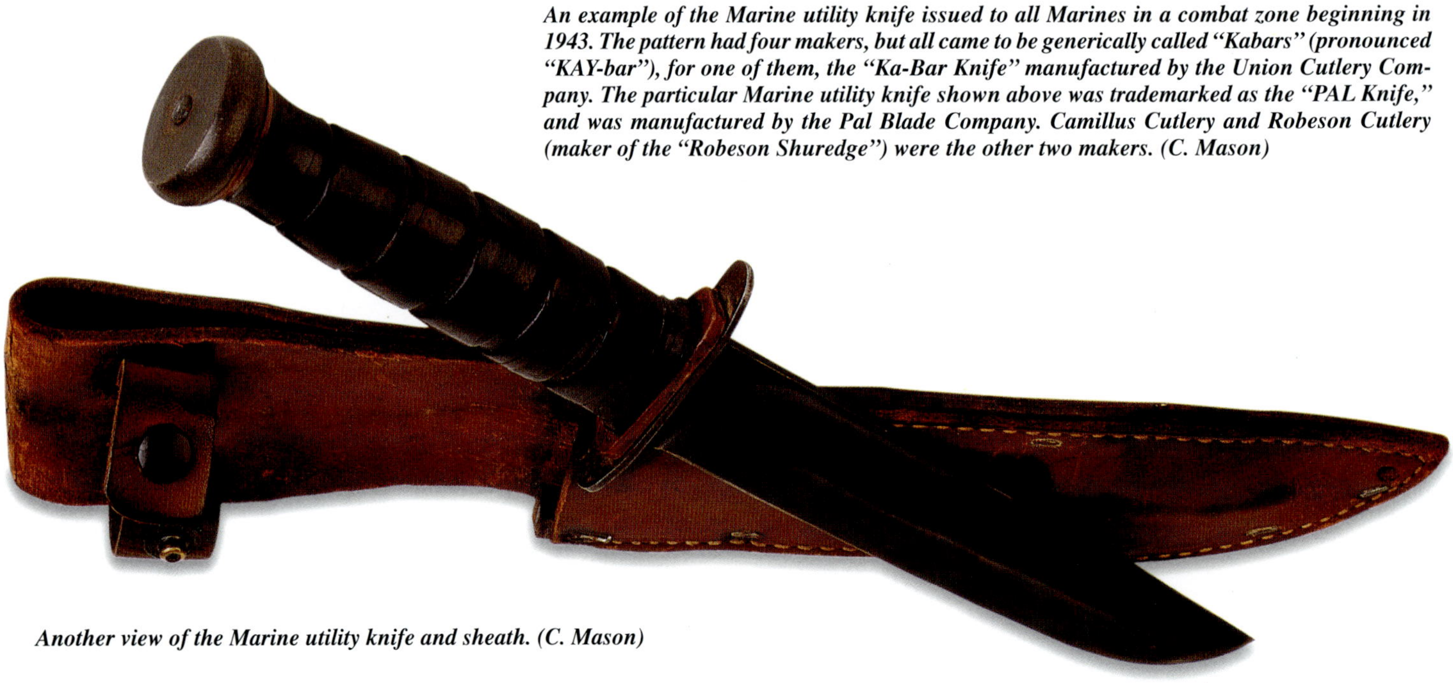

Another view of the Marine utility knife and sheath. (C. Mason)

The USMC stiletto made only by the Camillus Cutlery Company of Camillus, New York. It was modeled after the British commando knife. The knife was issued to Raiders and Paramarines but, in spite of being solid steel, was delicate, did not stand up well under field conditions, was not very useful other than as a weapon (i.e., for whittling stakes or opening cans) and thus was not overly popular with the troops. There were three types of sheaths, with minor variations and with and without metal reinforcement at the bottom. (C. Mason)

A close-up view of the blade angled to show the faint and delicate USMC scroll etching, which is difficult to photograph. (C. Mason)

A detail of the blade showing the Camillus markings. (C. Mason)

An angled view of the stiletto showing the crossguard. These crossguards are often found cracked and broken today. 14,370 of these knives were made during the war according to Camillus records. (C. Mason)

400

Left: Marines on Kwajelein Atoll interrogate a rare Japanese prisoner under close scrutiny from several Marines holding rifles at the ready. (Japanese soldiers rarely allowed themselves to be captured, and often kept pistols or hand grenades hidden to kill themselves and their captors in the event they were taken alive.) The Marine squatting in the foreground with the camouflage helmet cover is wearing a USMC stiletto on his right hip. (USMC)

Below: Marine forward observers on Iwo Jima. The Marine closest to the camera wears a "Kabar" on his right hip.

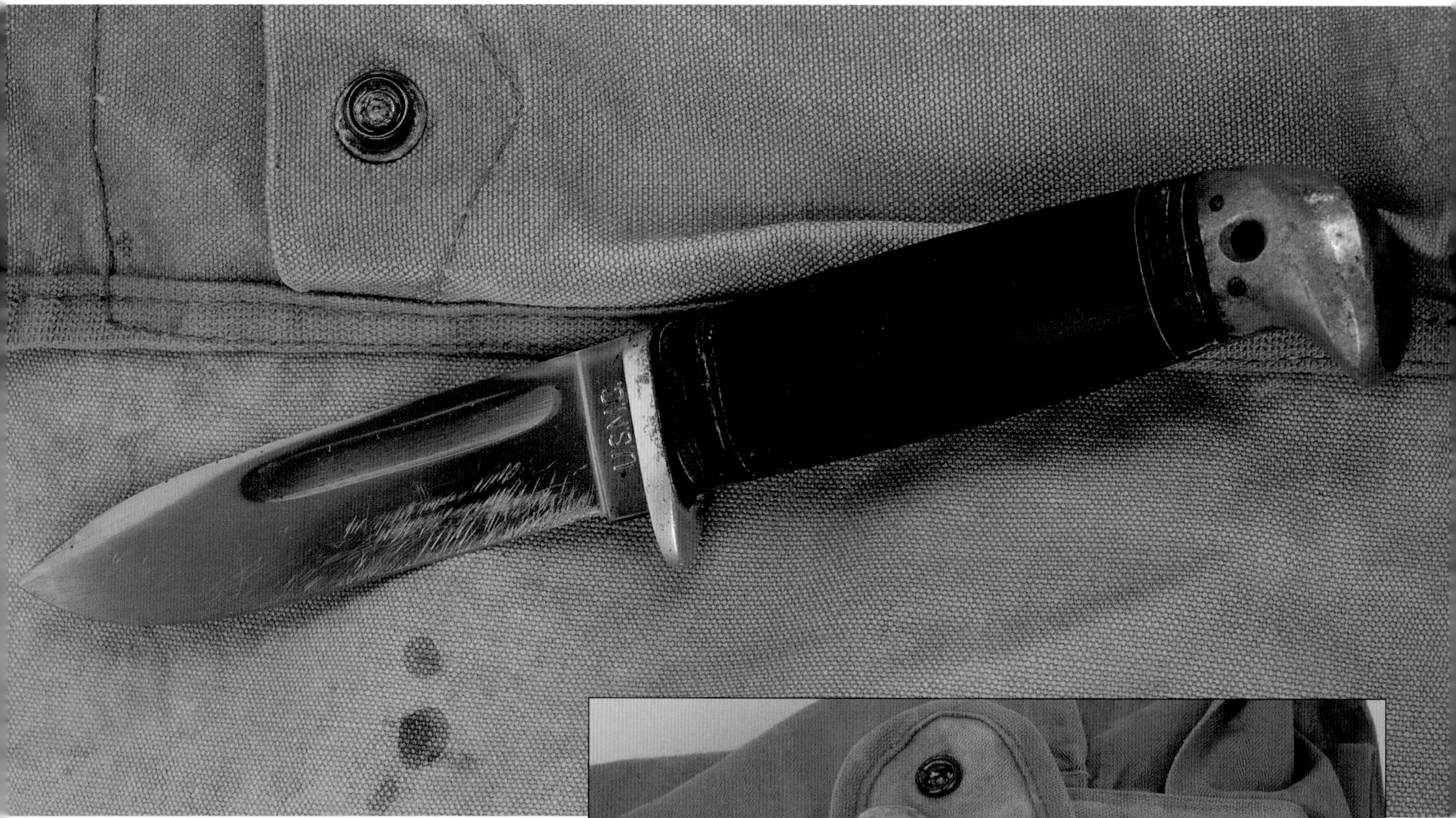

Above: The rarest of all US military fighting knives of World War II is the Western W31 Marine Parachutist knife made by the Western States Cutlery Company. Most collectors today agree that as few as 1,000 were ever produced. Only about a dozen are accounted for today. (C. Mason)

Right: The Marine Parachutist knife was only intended for issue at USMC parachute schools, and then only for the duration of the training course. It was to be turned back in to the Supply Officer at the completion of training. Some of the surviving examples are locally serial-numbered (probably at one of the parachute schools), others are not. (C. Mason)

The knife was to be carried on parachute jumps and used by the jumper to cut himself free of tangled shroud lines if caught in a tree or dragged across the drop zone by a high wind. A lanyard hole in the butt allowed the jumper to secure the knife to his equipment. The knife was issued with a brown leather sheath for wear on a belt and could also be carried in the knife pocket of the 1st model jump smock. (C. Mason)

In this rather dramatic photo, paramarine Francis R. Gildea carries a folding stock Reising and carries his Western W31 parachutist knife in his teeth. Gildea served in Company L. 3rd Marine Parachute Battalion. (USMC vla G. Petersen)

Above: Two Paramarines field stow a parachute after a practice jump at Camp Gillespie. Both are carrying W31 Marine parachutist knives in the right rear pocket of their coveralls. Both knives appear to be secured by lanyards. The man at left is also wearing an Army-style airborne helmet liner. (USMC via G. Petersen)

Right: The Paramarine in the foreground in this photo (taken in training at the jump school in San Diego) is wearing the W31 USMC Parachutist Knife in its leather sheath on his web belt. (USMC)

Below: The Paramarines in the photo are inspecting gear for a jump. The man closest to the camera is carrying his W31 knife in its sheath his smock pocket. The outline can be clearly seen. (USMC)

ACHUTISTS PASS
S AGAIN
USMC

Another W31 Marine Parachutist knife. Left: A detail of the USMC marking on the blade of the W31 Marine Parachutist knife. (L. Jacobsen)

A custom-made fighting knife carried by Paramarine Private First Class Walter J. Uhlman, Company C, 1st Marine Parachute Battalion. Such fighting knives were popular with front line troops and are often seen in period photos. (L. Jacobsen)

Above: The M1911A1 Colt .45 ACP semi-automatic pistol, here shown with magazine removed, together with the leather issue holster. Note that an officer's metal USMC EGA emblem has been affixed to the leather flap of the holster. (J. Busardo)

Right: The pistol could be carried in this waterproof issue bag for protection against the elements when not in the front lines. (J. Busardo)

The issue M-4 shoulder holster for the Colt .45 pistol. (J. Busardo)

Below: Paramarine First Lieutenant Pat W. Densman in training with the M1911A1 pistol. He carries his Reising SMG slung over his shoulder, wears the first model parachutist coverall (smock), and has a canvas pistol holster. Densman served in Company C, 1st Marine Parachute Battalion and fought on Gavutu. (USMC)

Above: The USMC issue lanyard for the Colt .45 pistol. (J. Busardo)

Right: A magazine and USMC-marked pouches for the .45. (J. Busardo)

One of the most common weapons issued to Paramarines was the .45 caliber Reising M55 submachine gun with folding stock. Because of its compact size, it was originally believed to be an ideal weapon for parachutists, but it tended to jam easily under sandy conditions and was thus very unpopular with the troops. (J. Busardo)

The left side of the M55 Reising with stock extended. (J. Busardo)

The right side with the stock folded. (J. Busardo)

The left side with the stock folded. (J. Busardo)

Above: A group of Paramarines from Company A and Headquarters Company, 1st Marine Parachute Battalion, under fire on Bougainville and awaiting evacuation by sea. The man closest to the camera carries one of the later M55 Reisings without the muzzle compensator. The man on his back with an M1911 Colt .45 caliber pistol is Private First Class Wallace H. Ford, Jr, of Headquarters Company. (USMC)

Below: A detailed view of the Reising markings. (J. Busardo)

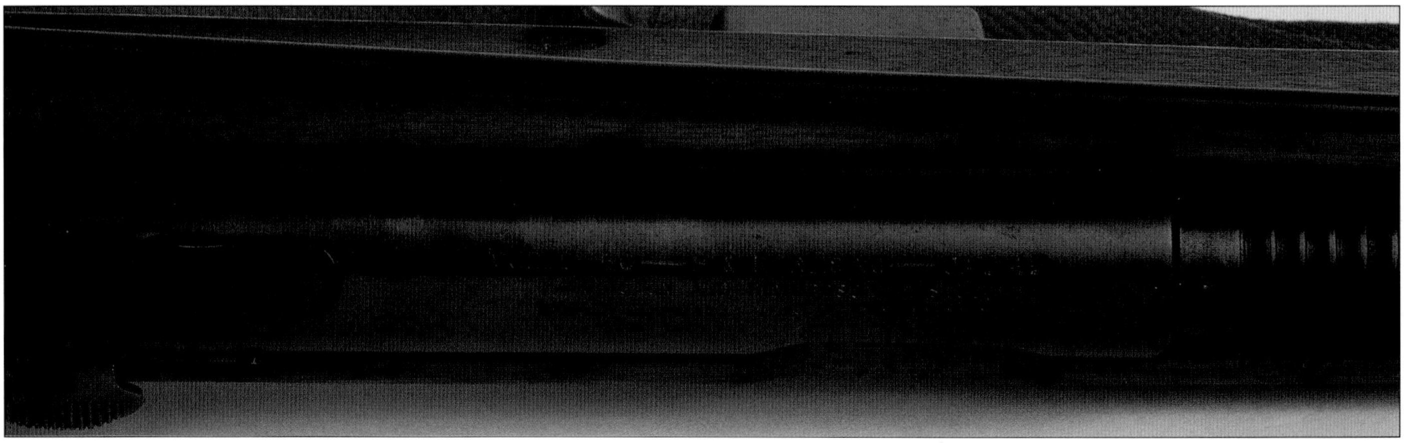

Right: The Reising could be carried on jumps in this camouflage bag. Doctrine called for it to be carried on top of the chest reserve parachute. (J. Busardo)

Below: The Reising could also be carried on jumps in this padded camouflage bag. (J.Busardo)

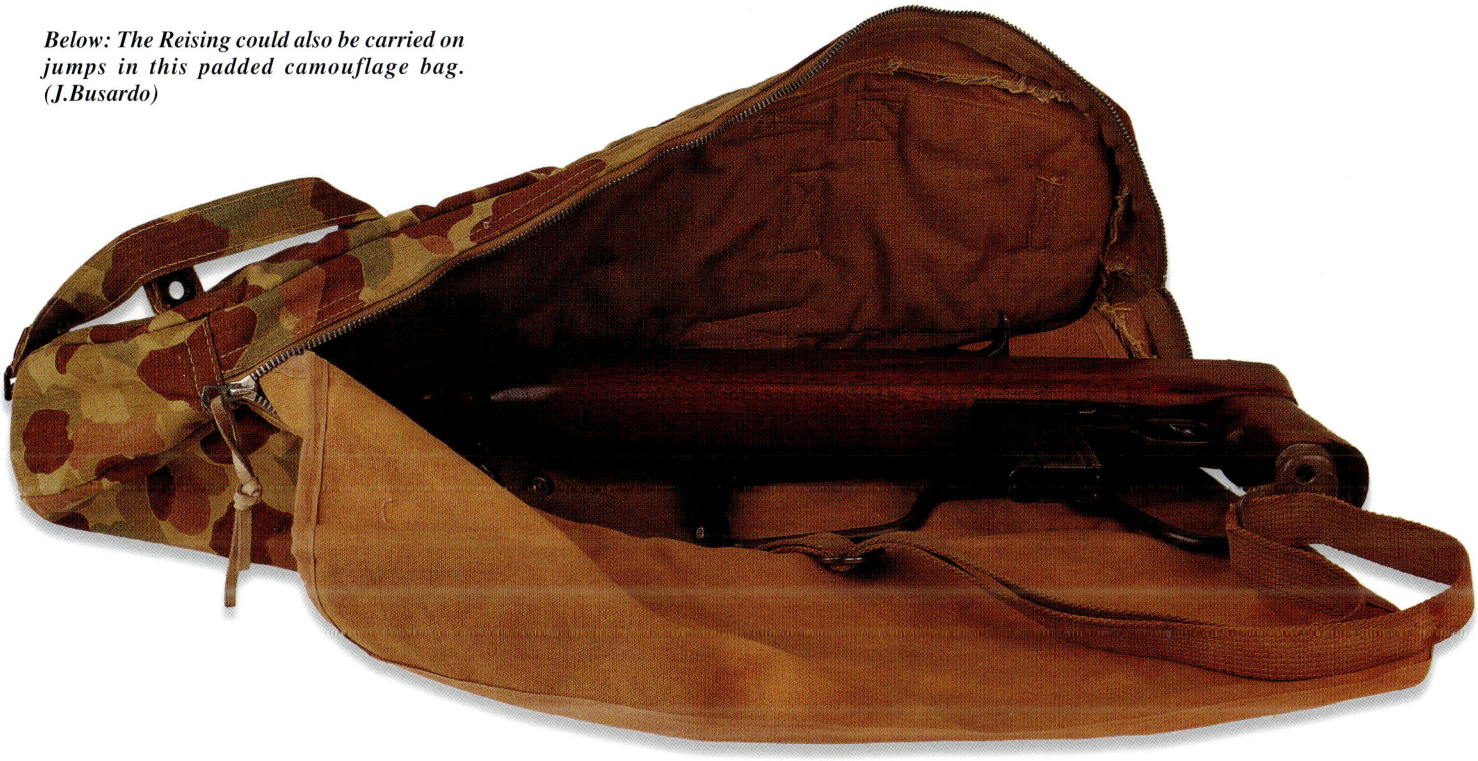

Below: The front of the drop bag. Bottom: The reverse of the padded drop bag. (J. Busardo)

Below: The Reising was also produced in a fixed stock version, sometimes used for training in the U.S. Right: A detail of the Reising and USMC Reising sling. (J. Busardo)

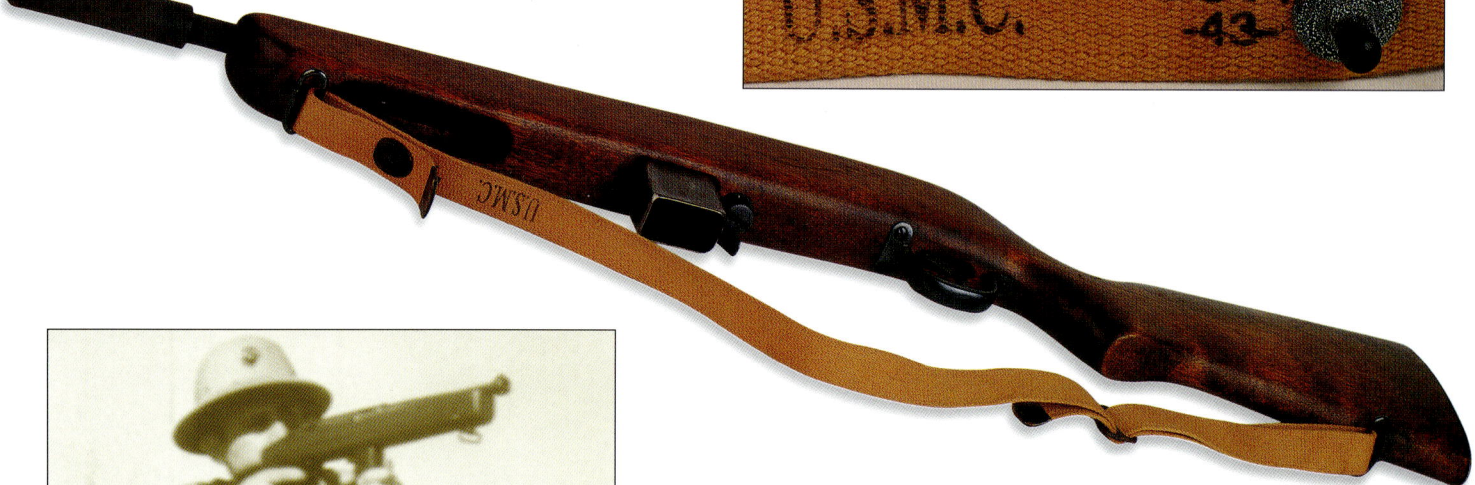

Opposite: The Paramarine on the cover of this issue of Mechanix Illustrated carries his Reising ready to jump without either camouflage case. The Reising he carries is one of the first 10,000 delivered to the Marine Corps. The early weapons were converted M50 commercial weapons which retained their blued finish and muzzle compensators. Later weapons were Parkerized and had no compensators.

Left:Paramarine Sergeant Edward Gruber at the firing range qualifying with the M55 Reising. (C. Mason)

Below: A group of Paramarine officers on Tulagi after the seizure of Gavutu, the first combat action of the Marine parachute forces. Several, including the Marine in the center of the back row, are carrying M55 Reisings. Note that they are all wearing the one piece USMC sage green coveralls used in Paramarine training. The jump boots on the officer in the front row, fourth from left, are clearly visible. (USMC)

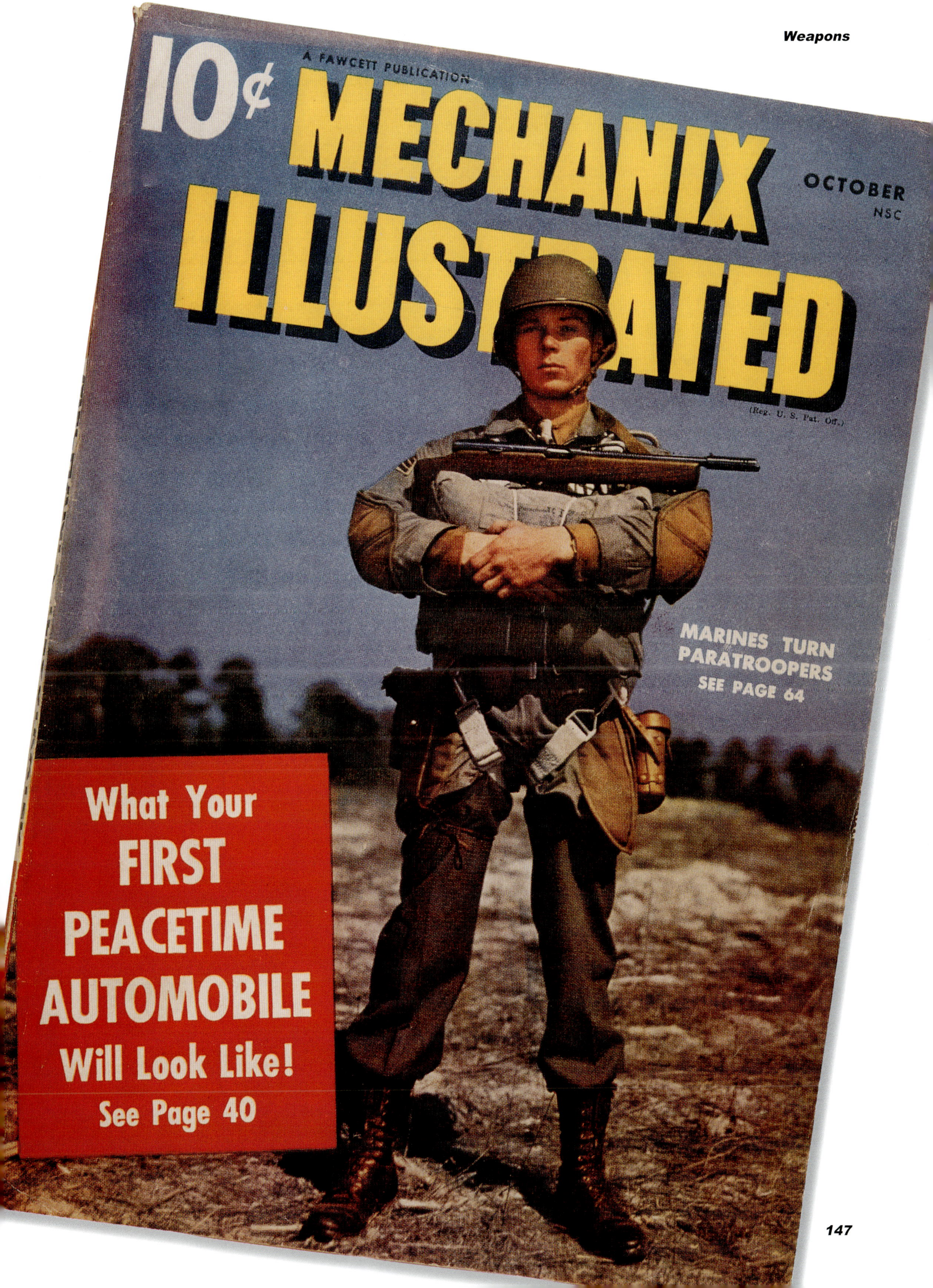
10¢
A FAWCETT PUBLICATION
MECHANIX ILLUSTRATED
OCTOBER
NSC
(Reg. U. S. Pat. Off.)
MARINES TURN PARATROOPERS
SEE PAGE 64
What Your FIRST PEACETIME AUTOMOBILE Will Look Like!
See Page 40

Another common weapon carried by the Paramarines was the M1A1 folding stock carbine. (J. Busardo)

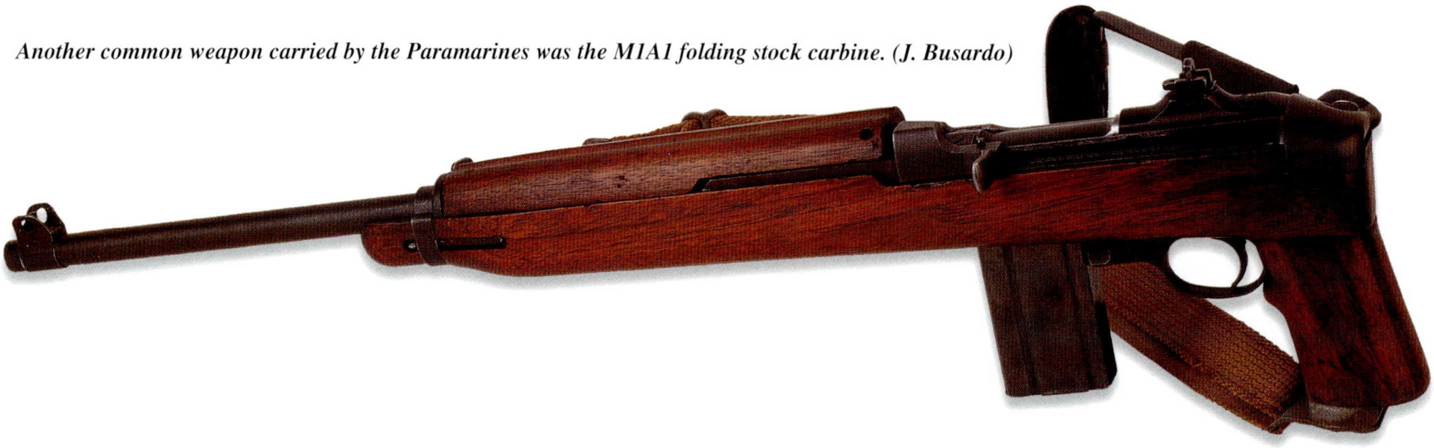

The M1A1 carbine with the stock extended. (J. Busardo)

A view of the steel buttplate. (J. Busardo)

A detailed view of the receiver. (J. Busardo)

The carbine could also be carried in the camouflage bag. (J. Busardo)

The USMC carbine magazine pouch with magazine. (J. Busardo)

The back side of the carbine magazine pouch. (J. Busardo)

Left: Three Paramarines at Camp Gillespie. The men on the left and right carry folding stock carbines, while the Marine in the center carries the M55 folding stock Reising. (USMC)

Below: A Paramarine in training at Camp Gillespie, San Diego, takes aim with his folding stock carbine. (USMC)

Below, top to bottom (J. Busardo):
The camouflage drop bag, front side, with flap closed.

The camouflage drop bag, front side, with flap open.

The interior of the camouflage drop bag.

The reverse of the bag showing the belt or equipment loop.

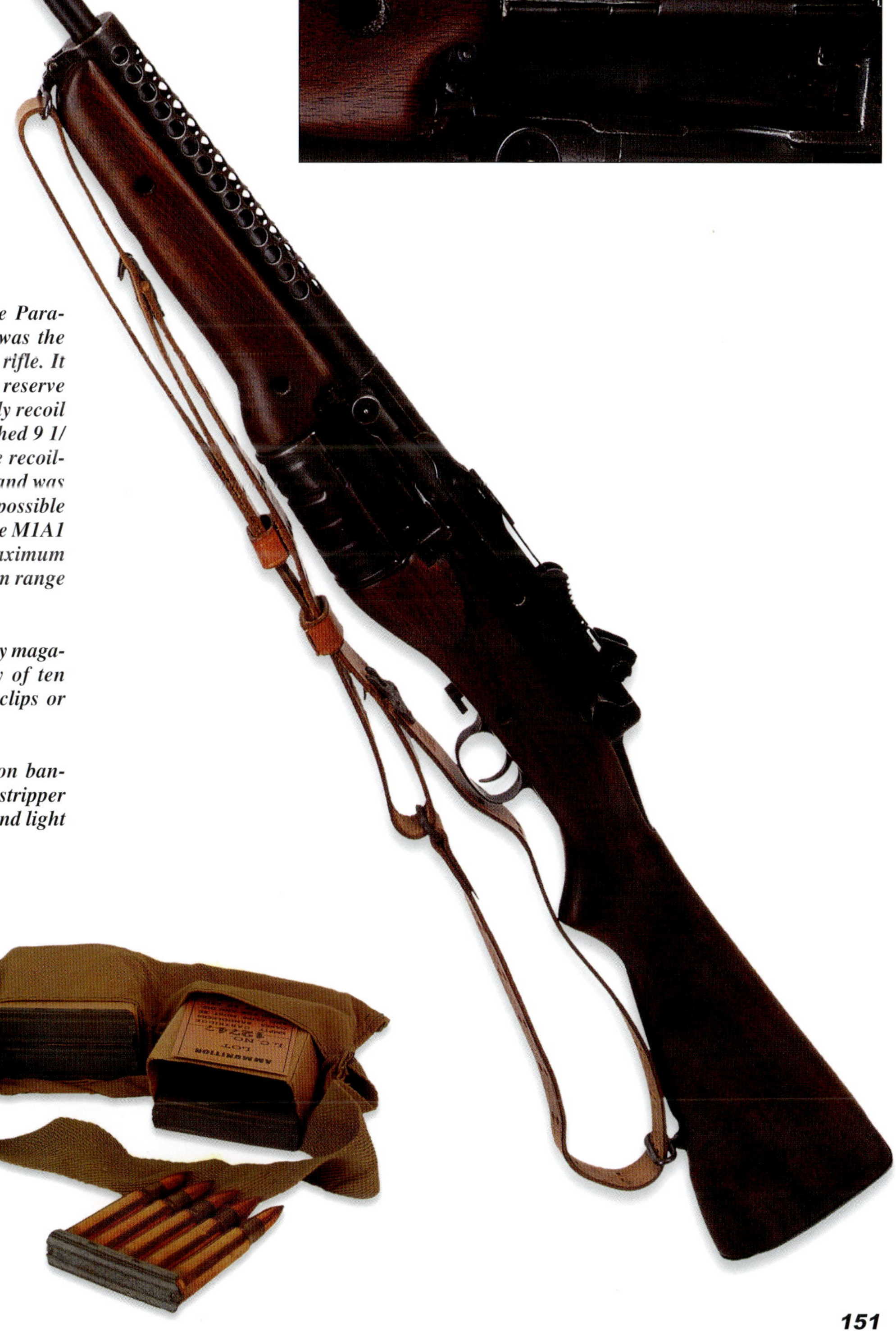

The most interesting weapon issued to the Paramarines, which was unique to that force, was the M1941 Johnson .30 caliber semiautomatic rifle. It was developed by Melvin Johnson, a Marine reserve officer. The rifle was recoil operated, the only recoil operated rifle used in World War II. It weighed 9 1/2 pounds and was 45 7/10 inches long. The recoiling Johnson barrel had four rifled grooves and was easily removed from the weapon, making it possible to carry it in the camouflage drop bag like the M1A1 carbine and the M55 Reising. It had a maximum effective range of 600 yards, and a maximum range of 3500 yards. (J. Busardo)

Top right: A close-up view of the unique rotary magazine. The rotary magazine had a capacity of ten rounds and could be loaded with stripper clips or with single rounds. (J. Busardo)

Below: The 6-pocket disposable ammunition bandoleer containing 60 rounds in five round stripper clips of the type used for the Johnson rifle and light machine gun. (J. Busardo)

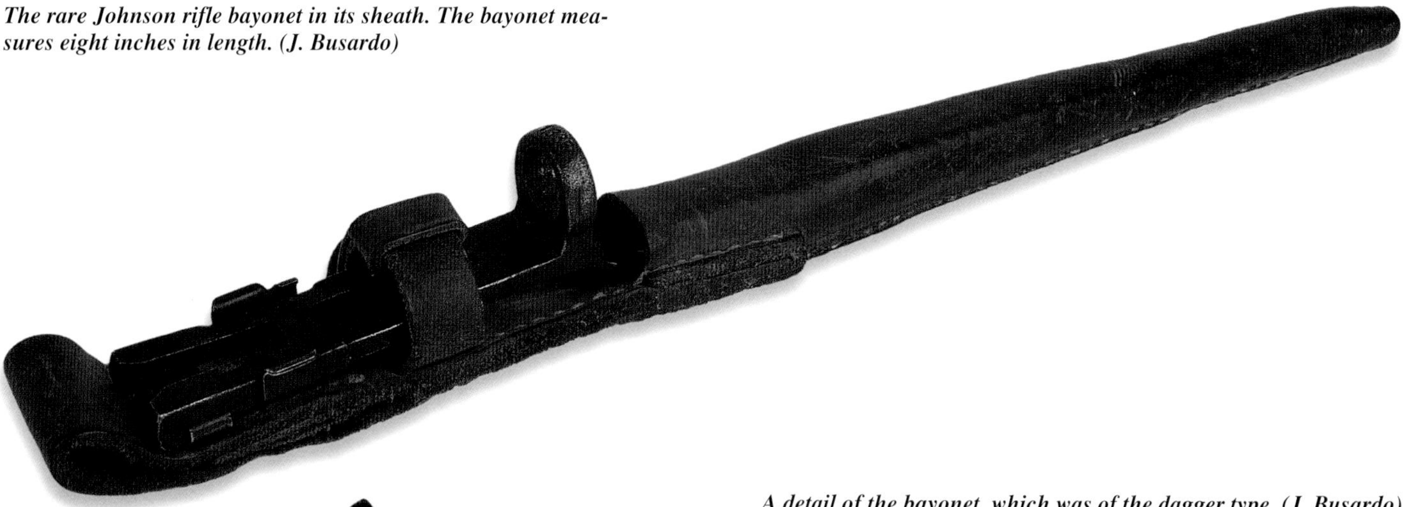

The rare Johnson rifle bayonet in its sheath. The bayonet measures eight inches in length. (J. Busardo)

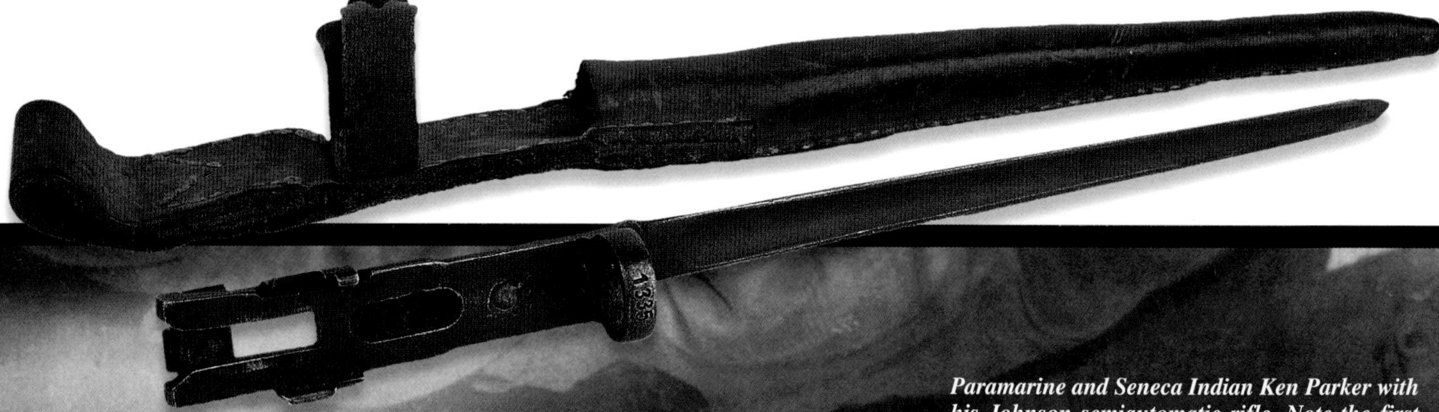

A detail of the bayonet, which was of the dagger type. (J. Busardo)

Paramarine and Seneca Indian Ken Parker with his Johnson semiautomatic rifle. Note the first model smock hanging in the background on the tent wall, the jump boots he is wearing and the spare pair under his cot. (J. Skutt)

A wartime patriotic poster for the home front. The weapon on the beach is clearly a Johnson LMG with bipod, rendered with some artistic license. (L. Jacobsen)

Additional firepower at the squad level was provided by the .30 caliber M1941 Johnson light machine gun. This view of the left side shows the curved magazine. This particular Johnson LMG, serial number 69, was issued to the Marine Corps in 1942. (J. Busardo)

A view of the Johnson LMG full length with the bipod folded and the rare sling attached. There were three models of bipod. (J. Busardo)

The right side of the Johnson LMG showing the ejection port. (J. Busardo)

Paramarine Corporal Neil Swanson training with the Johnson LMG. Note the sling and the third model bipod as seen in the pictured example. (USMC)

Right: Another photograph from training showing the curved magazine. (USMC)

The M9A1 bazooka. (J. Busardo)

A detail of the reinforcing webbing required on the inside of the bazooka carrying bag. (J. Busardo)

The bazooka shown folded with a 2.36 inch high explosive rocket. (J. Busardo)

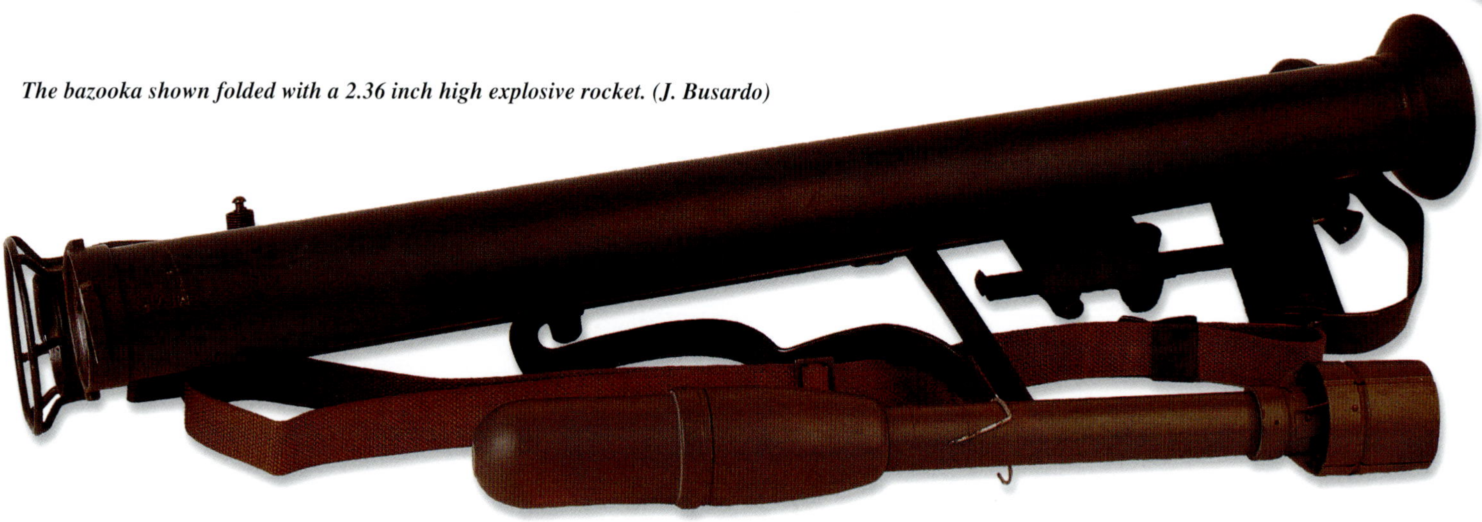

The bazooka could be carried and jumped with using this camouflage carrying bag. (J. Busardo)

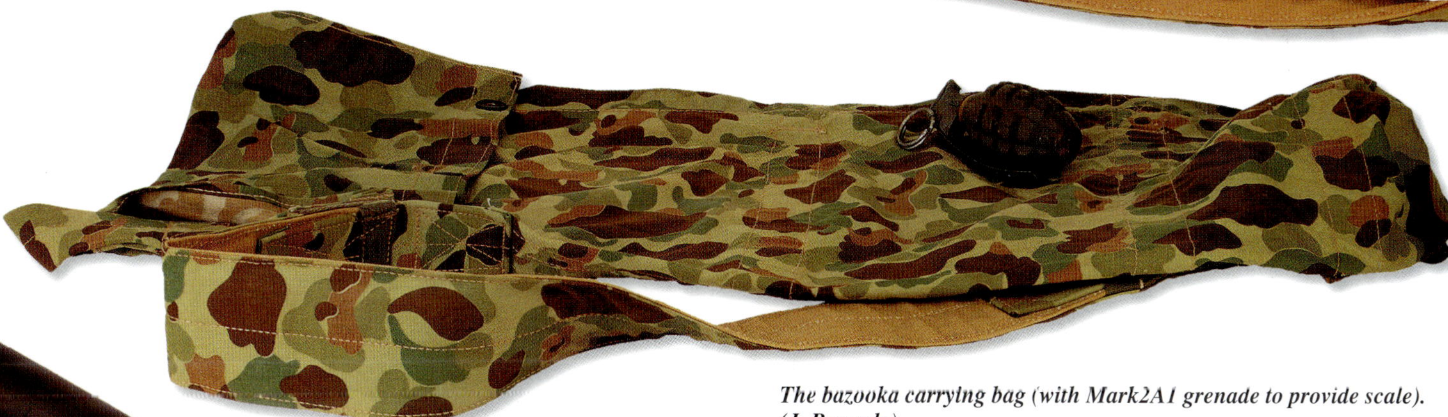

The bazooka carrying bag (with Mark2A1 grenade to provide scale). (J. Busardo)

Bazooka rockets had their own carrying bag. (J. Busardo)

The rockets came protected and packed in individual cardboard tubes. Up to three rockets could be carried in one bag. (J. Busardo)

Above: A full platoon of Paramarines stands inspection with weapons. The man in the rear rank, third from right, carries the M1919A4 .30 caliber Browning machine gun. The Paramarine next to him carries the M2 tripod. (USMC)

Opposite
Top: A Marine bazooka team in action on Pelilieu armed with the M1A1 bazooka. Note the spare rockets and cardboard tubes on the ground in the foreground. (USMC)

Center right: A group of Paramarines in training, probably at Camp Gillespie, California. The Marine on the far left is carrying an M1918 A2 Browning Automatic Rifle (BAR). (E. Queen)

Bottom: A famous image from the savage fighting on Okinawa, May 1945. One Marine takes aim with his M1 Thompson submachine gun, while his buddy with a model 1918A2 BAR ducks away from the line of fire. Marines often complained that the Thompson was heavy and required constant oiling to prevent rusting in the humid Pacific climate. (USMC)

The five-pocket Thompson magazine pouch. (J. Busardo)

The reverse of the Thompson five-pocket pouch, marked USMC. (J. Busardo)

The front of the Thompson five-pocket magazine pouch. This one was manufactured by R.M. Company in 1942. (J. Busardo)

The 20-round Thompson magazine. (J. Busardo)

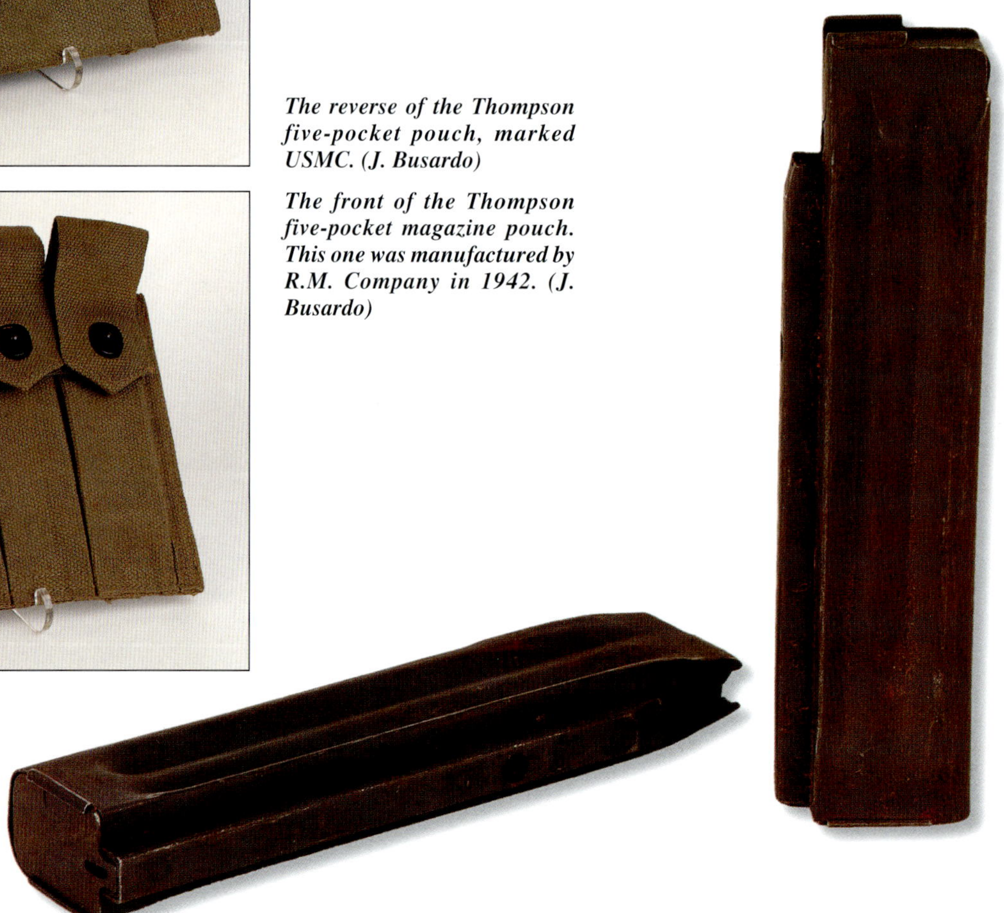

The ubiquitous "pineapple grenade," designated the Mk2A1.(J. Busardo)

A Paramarine in training prepares to throw a training grenade. (USMC via G. Petersen)

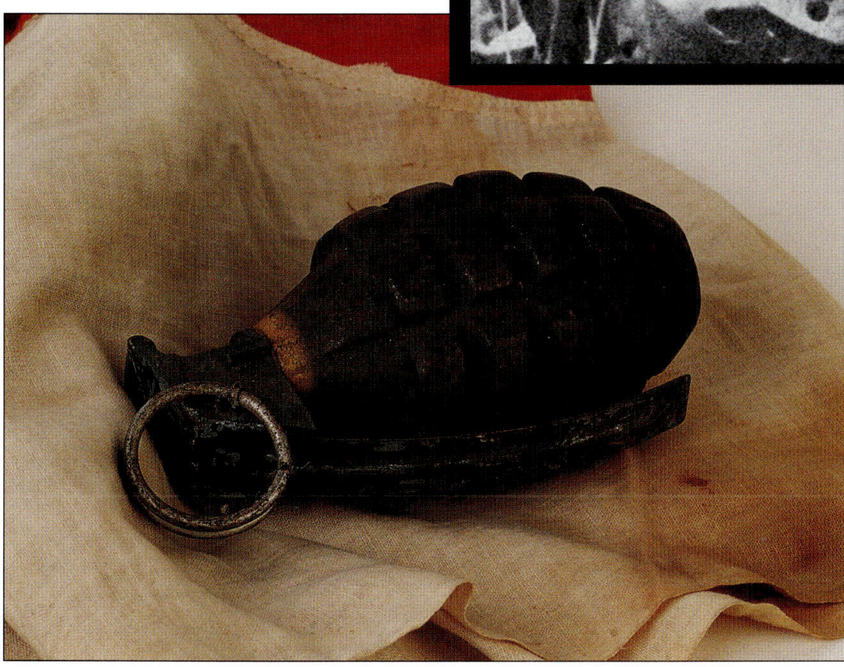

Another view of the Mk2A1 grenade showing the activating spoon and pin. (J. Busardo)

A carefully posed publicity shot of four Paramarines early in the program, two with reserve parachutes and two with Thompson submachine guns with rotary drum magazines. The man second from left has his parachute harness on inside out; the leg clip on his right leg is supposed to come up from the crotch as the other the Paramarines are wearing it, not down. (USMC)

A group of well-armed Paramarines in the Pacific pose with their weapons for a light-hearted photo. Three are carrying the Johnson LMG (J. Busardo)

CHAPTER FIVE

Wings and Insignia

Paramarine Lieutenant Hersel Blasingame wearing basic parachutist wings. The wings adopted by the Marine Corps were the same as the Army silver design, but tend to be much darker metal. The term "silver wings" is used within the Marine Corps to denote the Army pattern parachute wings (earned after five jumps at jump school) as opposed to Navy and Marine Corps "gold wings," and that convention is observed here. (C. Mason)

The "silver jump wings" which were worn by the late Paramarine Private First Class Warren G. McGuire of Eugene, Oregon. (C. Mason)

The reverse of Private First Class McGuire's jump wings. McGuire served in the 1st Marine Parachute Regiment. (C. Mason)

The silver jump wings worn by Paramarine Sergeant Harold A. Hoeschen. (L. Jacobsen)

The reverse of Sergeant Hoeschen's jump wings. (L. Jacobsen)

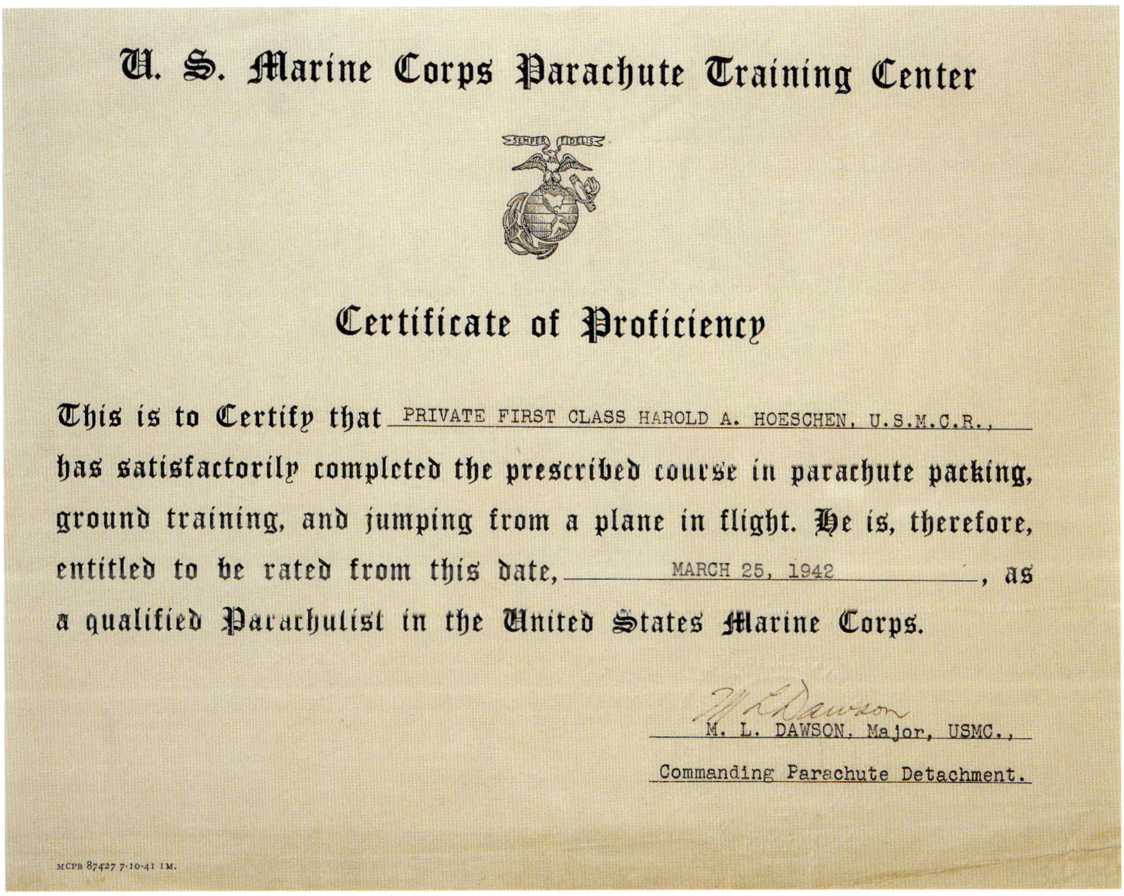

U. S. Marine Corps Parachute Training Center

Certificate of Proficiency

This is to Certify that PRIVATE FIRST CLASS HAROLD A. HOESCHEN, U.S.M.C.R., has satisfactorily completed the prescribed course in parachute packing, ground training, and jumping from a plane in flight. He is, therefore, entitled to be rated from this date, MARCH 25, 1942, as a qualified Parachutist in the United States Marine Corps.

W L Dawson
M. L. DAWSON, Major, USMC.,
Commanding Parachute Detachment.

MCPB 87427 7-10-41 IM.

The Certificate of Proficiency as a Marine Parachutist earned by Sergeant Harold A. Hoeschen, Company C, 1st Marine Parachute Battalion, entitling him to wear the coveted silver wings of the paratrooper. (L. Jacobsen)

A second set of wings owned by Paramarine Sergeant Harold A. Hoeschen. (L. Jacobsen)

The reverse of these wings. (L. Jacobsen)

Non-regulation Paramarine jump wings with a miniature USMC Eagle, Globe and Anchor (EGA) insignia affixed to the shroud lines. These wings were worn by Paramarine Erwin Albers, Weapons Company, 1st Marine Parachute Regiment. Right: The reverse of these wings. (L. Jacobsen)

Paramarine jump wings worn by Private First Class Edward L. McClure, F Company, 2nd Marine Parachute Battalion. Right: The reverse of McClure's jump wings. (C. Mason)

U.S. Marine Corps Parachute Training School

SEMPER FIDELIS

Certificate of Proficiency

This is to certify that PRIVATE FIRST CLASS (SS) CHARLES C. CLOW, Jr., U.S.M.C. has satisfactorly completed the prescribed course in parachute packing, ground training, and jumping from a plane in flight. He is, therefore, entitled to be rated from this date, 28 April, 1943, as a qualified Parachutist in the United States Marine Corps.

C.E. Shepard Jr.
C.E. SHEPARD, Jr.,
Lieutenant Colonel, U.S. Marine Corps.

-16—1141—M.I.N.Y. 11-23-42—5M

The Certificate of Proficiency as a Marine Parachutist earned by Corporal Charles C. Clow, Jr. (a Private First Class at the time). (L. Jacobsen)

A rare set of jump wings of the dark USMC style made by the firm of Angus & Coote of Sydney, Australia. Right: The reverse of the Angus & Coote wings, showing the markings. (L. Jacobsen)

Major Harry L. Torgerson on Bougainville in 1943. He is wearing jump wings on his khaki shirt. His unique carbine version of the Johnson semiautomatic rifle is next to him. Torgerson fought on Gavutu, where he earned the Silver Star. He took command of 1st Marine Parachute Battalion on Bloody Ridge (Guadalcanal) and later took command of the 3rd Marine Parachute Battalion on Bougainville when the commanding officer was wounded. (USMC)

Corporal Hermann C. Glass wearing the silver parachutist wings. Note the EGA device affixed to the Expert Rifleman Badge. (E. Queen)

Below: Two Marine Paratroopers out on the town with their girls. On the left is Corporal Hermann C. Glass. Corporal Glass is wearing the "Paramarine patch," on his left shoulder; his unidentified buddy on the right is not. (E. Queen)

Left: The jump wings which belonged to Paramarine Louis Stanley Hardt, Company L, 3rd Marine Parachute Battalion. Right: The reverse of Hardt's jump wings, on which he has engraved his name. (C. Mason)

Below: The Certificate of Proficiency as a Marine Parachutist earned by Paramarine Private Waldemar A. Neudeck. Neudeck was wounded in action on Iwo Jima with 27th Marines and earned the Silver Star. (L. Jacobsen)

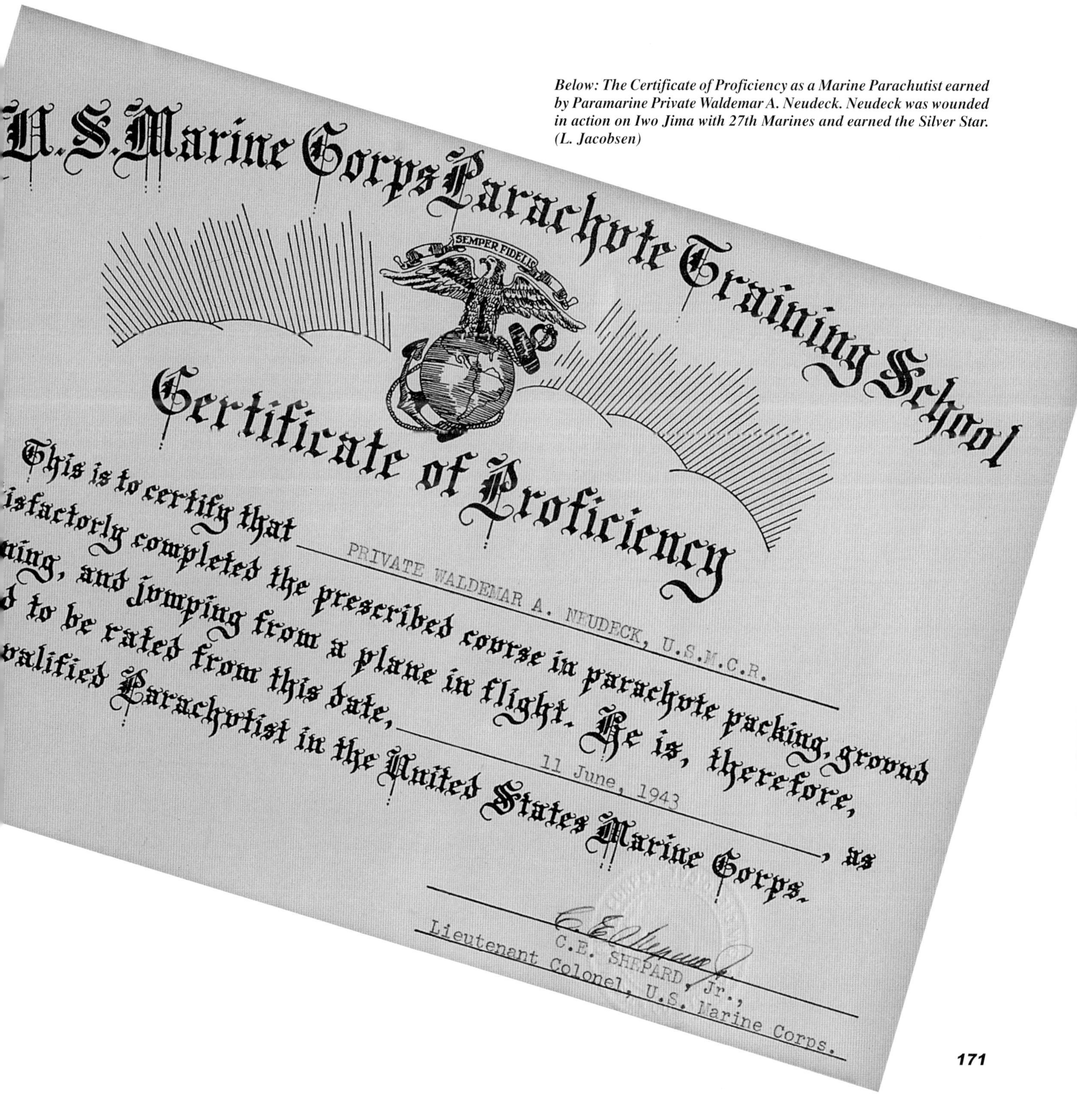

U.S. Marine Corps Parachute Training School

SEMPER FIDELIS

Certificate of Proficiency

This is to certify that PRIVATE WALDEMAR A. NEUDECK, U.S.M.C.R.
...isfactorly completed the prescribed course in parachute packing, ground
...ning, and jumping from a plane in flight. He is, therefore,
...d to be rated from this date, 11 June, 1943, as
...ualified Parachutist in the United States Marine Corps.

C.E. SHEPARD, Jr.,
Lieutenant Colonel, U.S. Marine Corps.

Above: The jump wings, ribbons and qualification bars of Paramarine Corporal Edward G. Gruber, 1st Marine Parachute Regiment. (C. Mason)

Right: Corporal Gruber before attending Parachute School. (C. Mason)

Opposite
Top: The Certificate of Proficiency as a Marine Parachutist earned by Paramarine Private First Class Frederick H. Nicholls, Company B, 1st Marine Parachute Battalion.(L. Jacobsen)

Bottom: Front and reverse of a pair of unsold Paramarine jump wings made the Amico Company in their original box. The label inside the box lid of the Amico wings indicates they were labeled for sale in March 1945 by the firm of Henry Wosk of San Diego. (L. Jacobsen)

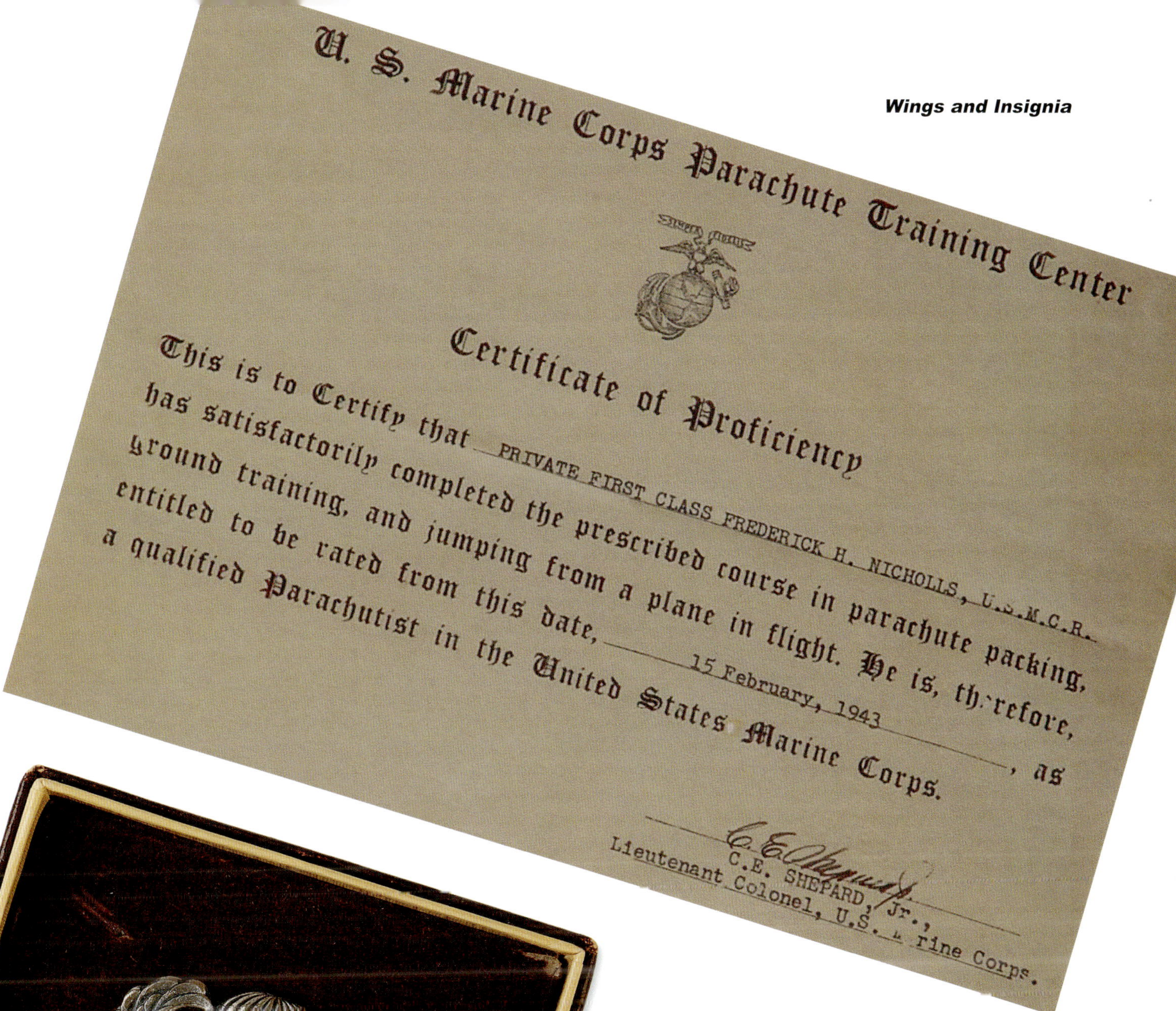
U. S. Marine Corps Parachute Training Center

Certificate of Proficiency

This is to Certify that PRIVATE FIRST CLASS FREDERICK H. NICHOLLS, U.S.M.C.R. has satisfactorily completed the prescribed course in parachute packing, ground training, and jumping from a plane in flight. He is, therefore, entitled to be rated from this date, 15 February, 1943, as a qualified Parachutist in the United States Marine Corps.

C.E. SHEPARD, Jr.,
Lieutenant Colonel, U.S. Marine Corps.

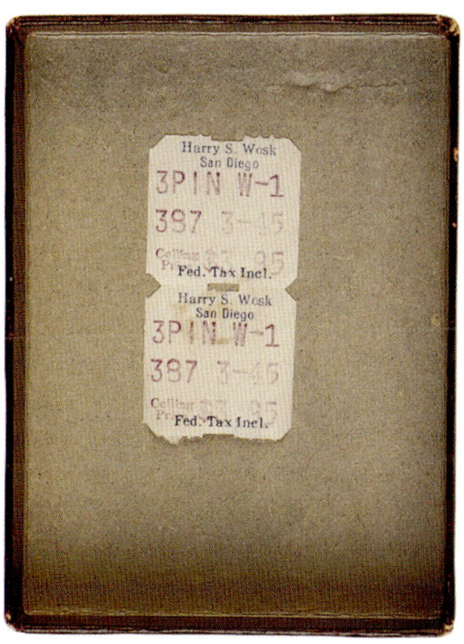

Above: Paramarine training differed from Army paratrooper training, most significantly in the fact that Marines and Navy personnel attending jump school learned to pack parachutes and packed all six parachutes they personally jumped with during training, a practice which continued until the last Navy Parachute School, at Lakehurst, New Jersey, was closed in the 1990's. Here Paramarine trainees under the watchful eye of a Rigger instructor in San Diego learn to lay out shroud lines before packing into a deployment bag. (USMC)

Individual Paramarines occasionally wore miniature jump wings on the right front of their overseas cap, a non-regulation practice to be sure, but one that reflected their pride in their status as paratroopers. Right: The reverse of the miniature wings. (C. Mason)

Paramarine Private First Class Gordon Pischke wearing the gold Navy and Marine Corps Rigger wings at Camp Gillespie, San Diego, California, in November 1943 (the photograph is dedicated on the back). Pischke was killed in action on Iwo Jima with 27th Marines. (L. Jacobsen)

Above: World War II gold Navy and Marine Corps Parachute Rigger Wings in their original box. These wings were originally intended to be worn by qualified parachute riggers as a replacement for the "Parachute Man" patch worn on the lower left sleeve. Photographic and anecdotal evidence however clearly show that the gold wings were often worn by Paramarines who, unauthorized and contrary to regulations, simply liked them better than the silver wings. After World War II, the gold wings were officially adopted in 1963 as "Navy and Marine Corps Master Parachutist Wings" for all personnel who completed 10 parachute jumps from an aircraft in flight while assigned to a billet which specifically required parachute duty and for which supplemental monthly parachute pay was authorized. Below: The reverse of the gold jump wings. (C. Mason)

The lid of the original box. Note the shopkeeper's penciled notation in the upper left corner of the box: "USMC Rigger." (C. Mason)

Above: An unidentified Navy Aviation Boatswains Mate wearing gold Rigger wings during World War II. (C. Mason)

Two at left: The front of the gold Rigger wings worn by Navy Parachute Rigger George J. Sherman during World War II. (L. Jacobsen)

Left: Rigger wings were also made in miniature size for the Mess Dress Uniform and occasionally were also worn, contrary to regulations, on the right hand side of the overseas cap. These wings were manufactured by the Amico firm. Below: The reverse of the miniature gold Rigger Wings. (C. Mason)

Below: Paramarine Corporal Charles C. Clow, Jr., (here a Private First Class) wearing the gold Navy and Marine Corps Rigger wings. Clow's complete Marine Corps service record indicates he did not attend Rigger School and therefore wore the gold wings seen here, like many Paramarines, as a matter of personal preference. (L. Jacobsen)

The lid of the Amico box for the miniature Rigger wings. (C. Mason)

Colonel Peter J. Ortiz, USMC, was one of several Marines to be assigned to the OSS in Europe for operations behind German lines. Ortiz served in the French Foreign Legion in the 1930's and spoke fluent French. He was wounded as a Legionairre in 1940, captured by the Germans, escaped, and returned to the US, where he joined the Marines. He was wounded again behind enemy lines in North Africa and was twice decorated with the Navy Cross. In this photograph he has just been awarded his second Navy Cross, the Legion of Merit and the Croix de Guerre by Colonel (later Major General) H. B. Liversedge. The Special Forces parachute wing is clearly visible on his left sleeve. (USMC)

Opposite
Corporal Hermann C. Glass, 1st Marine Parachute Regiment, wearing a very unusual set of embroidered gold parachutist wings on a darker cloth backing. He wears the patch of the 1st Marine Division on his left shoulder and miniature silver parachute wings on his khaki field scarf. (E. Queen)

Above: Members of Ortiz's OSS team behind German lines in the French Alps near Col-des-Saises render honors at the grave of Sergeant Charles R. Perry, USMC, a member of the team who died when his parachute failed to open. From the left they are: Major Ortiz, Captain Francis Coolidge, US Army Air Corps, Gunnery Sergeant Robert E. LaSalle USMC, Sergeant John P. Bodnar USMC, Sergeant Frederick Brunner USMC and Sergeant Jack Risler USMC. All the Marines are wearing their Special Forces jump wings on the left chests of their jackets. The Army officer is wearing his on his sleeve. Below: A close up of the same photograph shows the Special Forces wings more clearly. (USMC)

Sergeant Risler, left, and Sergeant Bodnar, right, were captured with Ortiz and other team members during the war and held as POWs. They are shown here in Paris after their release in 1945. Both are wearing their U.S. parachute wings on their green winter service coats. (USMC)

An example of the Special Forces (OSS) parachute wings as worn by Colonel Ortiz and other Marines in the unit. (D. Cone)

Left: The USMC Basic Badge, introduced in 1937, with a rare "Rigger" Qualification Bar. (L. Jacobsen)
Right: The reverse of the USMC Basic Badge with Rigger Bar showing the World War II period pinback device. (L. Jacobsen)

In this photograph, Corporal Hermann C. Glass wears the Fourragere on his Dress Blue Uniform. The Fourragere was awarded to the Fifth Marines and Sixth Marines during World War I, and these two regiments have been entitled to wear this unique distinction ever since. Because the Marine Corps typically eschews such adornments on their uniforms, the Fourragere is sometimes referred to disparagingly by other Marines as the "pogey rope." (E. Queen)

Left: The instruction manual for prospective Riggers at the Parachute Materiel School Lakehurst. This copy belonged to Paramarine Wilfred R. Downing, who has written his name in pencil on the cover. Note the facsimile of the Navy and Marine Corps Parachute Riggers Wings on the cover. Downing was assigned to the Air Delivery Section. (L. Jacobsen)

Right: The first page of the manual. Note the date: June 1, 1941. Presumably, this is the date the training manual contents were produced, and the cover with the gold parachute Rigger Wings insignia came later in the war. (L. Jacobsen)

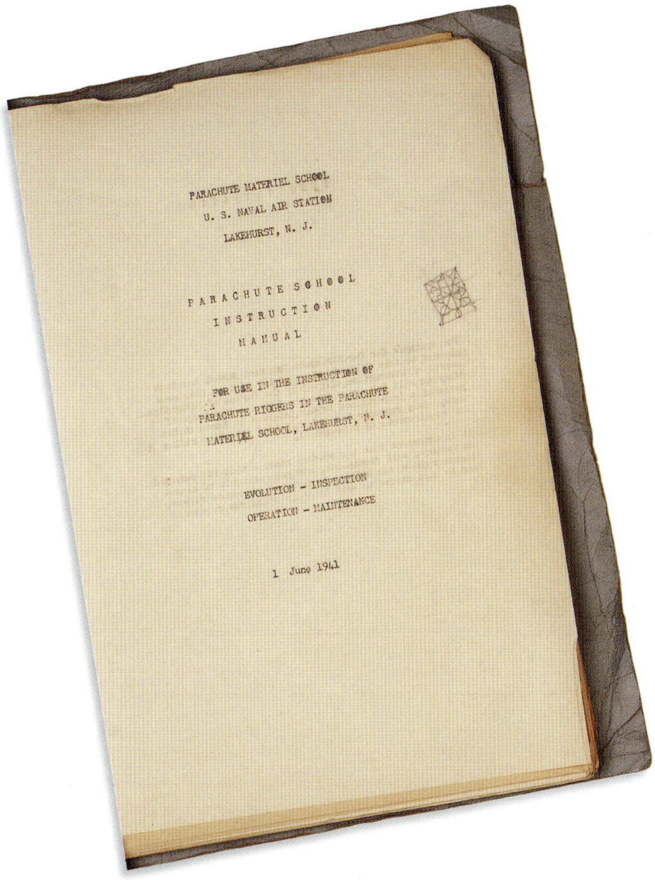

Below: The "Parachute Man" patch for qualified Riggers, which preceded the creation of the gold Navy and Marine Corps Parachute Rigger Insignia (wings). The patch was one of five qualification sleeve insignia authorized for Marines serving in ship's detachments. It clearly was worn more widely by USMC personnel however. Seen here is the original cellophane packet bearing the manufacturers label by Gemsco which the patch came in. This example, with red embroidery on forest green wool, was for wear on the lower left sleeve of the Marine Corps winter service coat. (L. Jacobsen)

Right: The insignia and jump wings of Paramarine Milton Peterson. Peterson was killed on Iwo Jima. (L. Jacobsen)

The "Parachute Man" insignia with gold embroidery on black wool backing for wear on the Marine Corps dress blue uniform. (L. Jacobsen)

The "Parachute Man" insignia with blue embroidery on gray cotton backing for wear on the Navy gray cotton service uniform. (L. Jacobsen)

The "Parachute Man" insignia with blue embroidery on white cotton backing for wear on the Navy dress white summer uniform jumper. (L. Jacobsen)

The "Parachute Man" insignia with white embroidery on blue wool for wear on the Navy dress blue winter uniform jumper. (L. Jacobsen)

Right: The "Parachute Man" insignia on the sleeve of the winter service uniform of Paramarine Walter F. "Fritz" Gemeinhardt, Air Delivery Section, 1st Marine Parachute Regiment. (Marine Corps Non-Commissioned Officer School, Parris Island, South Carolina)

Entire books have been written on the Eagle, Globe and Anchor (EGA) insignia of the United States Marine Corps and space does not permit a detailed examination of this insignia. The rarest of the World War II EGAs is shown here as being of special interest. It was manufactured by the firm of K C Luke of Melbourne, Australia. Right: The reverse of the KC Luke EGAs showing the characteristic pinback device of that firm. On some pairs, only one device is marked. On other pairs, both devices are found marked. Otherwise, virtually all EGAs produced during the war had screw post backings. (E. Wages)

Above: A USMC Paramarine Sweetheart Pin made by the Galawood-Roth firm of New York. (L. Jacobsen)

Left and below: A USMC Paramarine "Sweetheart Pin" made by the firm of AMICO. (L. Jacobsen)

Below: Another Paramarine Sweetheart Pin made by the Galawood-Roth firm of New York with the Marine Corps Bulldog mascot in place of the traditional EGA device. (L. Jacobsen)

Left: An unidentified Paramarine Corporal wearing the Australian-made 1st Marine Amphibious Corps Parachute Units shoulder patch of the J.J. Cash firm of Queensland. This patch is commonly referred to as simply the "Paramarine patch." (C. Mason)

Above and below: The Paramarine patch made by the J.J. Cash firm. If you ask three Paramarine veterans the meaning of the red border you'll hear three different answers. Some say it indicated a combat veteran. Others say it indicates the wearer was awarded the Purple Heart or that the wearer was a veteran of the Solomon Islands campaign. Yet another explanation is that it simply looked good. (C. Mason)

Left: This Paramarine patch made by the Australian firm of J.J. Cash was worn by Corporal Hermann C. Glass, 1st Marine Parachute Regiment. (E. Queen)

The Paramarine patch worn by Private First Class Frederick Nicholls. (L. Jacobsen)

Above right: The Paramarine patch made by the firm of Jocob Rhoner of Sydney. Right: The reverse of the Rhoner patch. (L. Jacobsen)

Left: The smaller U.S. made Para-marine patch, commonly called the "twill patch" in reference to its background. This is thought to be the only U.S.-made Paramarine patch actually worn before the 1st Marine Parachute Regiment was formally disbanded on February 29, 1944 (fittingly, Leap Day of Leap Year). Individual former Paramarines continued to purchase the patches privately and wear them afterwards in 1944 and 1945. Below: The reverse of the "twill patch." (L. Jacobsen)

Left: One of several slight variations of the large style World War II U.S.-made Paramarine patch. For comparison, note the "fat" border around the diamond shape. (L. Jacobsen)

Below: The reverse of this patch, which has been removed from a scrapbook or a display board. (L. Jacobsen)

Above: An unsold patch from the dead stock of the firm of Henry Wosk of San Diego. (L. Jacobsen)

Right: The reverse of the unsold Wosk patch, showing the original sales label. Note the date this patch was labeled for sale: January 1945. (L. Jacobsen)

Left: Another variation of the large U.S. made patch. On this one, the parachute is slightly smaller. (L. Jacobsen)

Below: The reverse of the "small parachute" variation. (L. Jacobsen)

Above: An unusual variation U.S.-made patch embroidered on a blue felt background. (L. Jacobsen)

Right: The reverse of the felt patch showing the cheesecloth backing. (L. Jacobsen)

Above: Corporal (later Sergeant) Art Sirles, Company B, 1st Marine Parachute Battalion, wearing the gold parachute wings and Paramarine shoulder patch. Corporal Sirles was not trained or qualified as a Rigger, and said he wore the gold wings out of personal preference. (D. Bailey)

Left: The design for the patch for the 3rd Marine Parachute Battalion designed by Walt Disney Studios and published in Leatherneck Magazine in July 1943. The author has never seen an original period example of this patch and it is highly unlikely that it was actually worn. (USMC)

Below: The Purple Heart awarded to Captain Gaylord S. Ziegler, USMCR for wounds received on Iwo Jima. Ziegler was a Paramarine who served in Headquarters Company, 3rd Marine Parachute Battalion. (L. Jacobsen)

Right: The award document which accompanied Ziegler's Purple Heart. It has been signed by the legendary Marine Corps General H.M. ("Howlin' Mad") Smith. (L. Jacobsen)

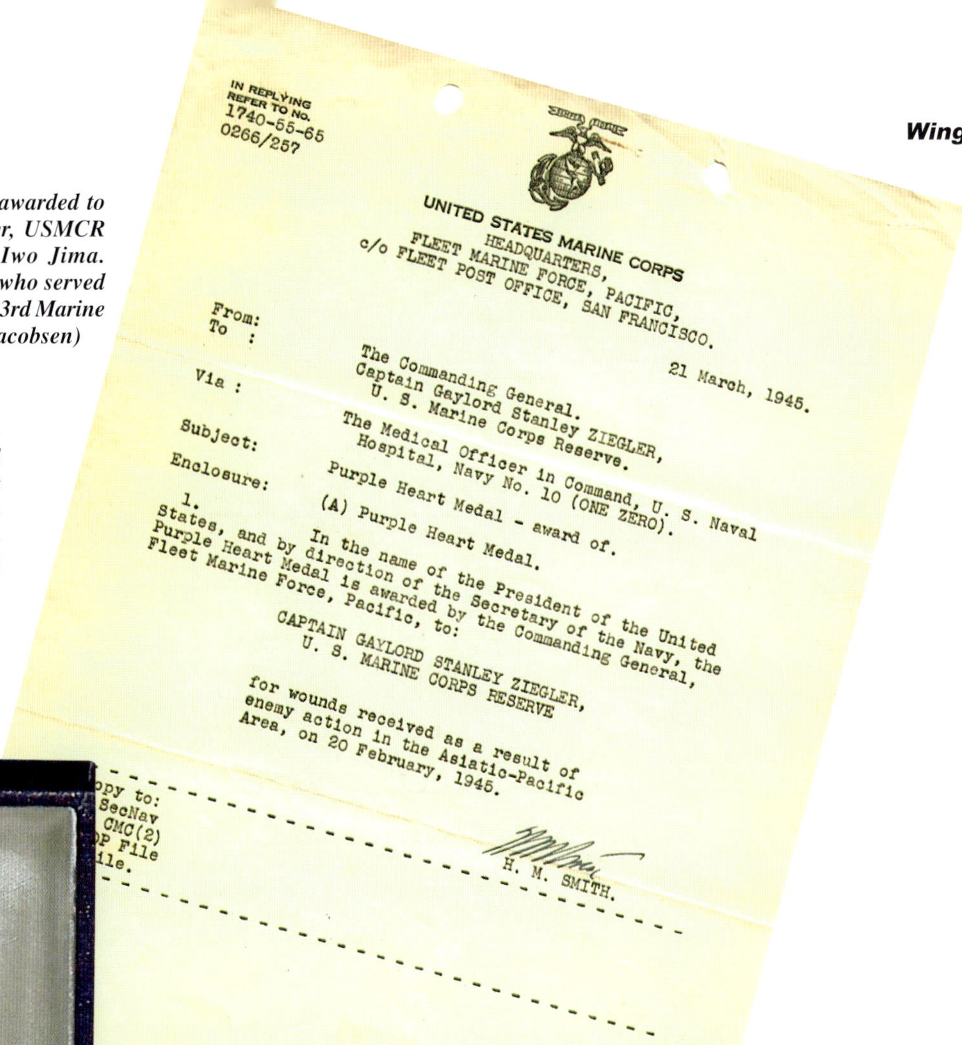

IN REPLYING REFER TO NO.
1740-55-65
0266/257

UNITED STATES MARINE CORPS
HEADQUARTERS,
FLEET MARINE FORCE, PACIFIC,
c/o FLEET POST OFFICE, SAN FRANCISCO.

21 March, 1945.

From: The Commanding General.
To : Captain Gaylord Stanley ZIEGLER, U. S. Marine Corps Reserve.
Via : The Medical Officer in Command, U. S. Naval Hospital, Navy No. 10 (ONE ZERO).
Subject: Purple Heart Medal - award of.
Enclosure: (A) Purple Heart Medal.

1. In the name of the President of the United States, and by direction of the Secretary of the Navy, the Purple Heart Medal is awarded by the Commanding General, Fleet Marine Force, Pacific, to:

CAPTAIN GAYLORD STANLEY ZIEGLER,
U. S. MARINE CORPS RESERVE

for wounds received as a result of enemy action in the Asiatic-Pacific Area, on 20 February, 1945.

H. M. SMITH.

Copy to:
SecNav
CMC(2)
...P File
...ile.

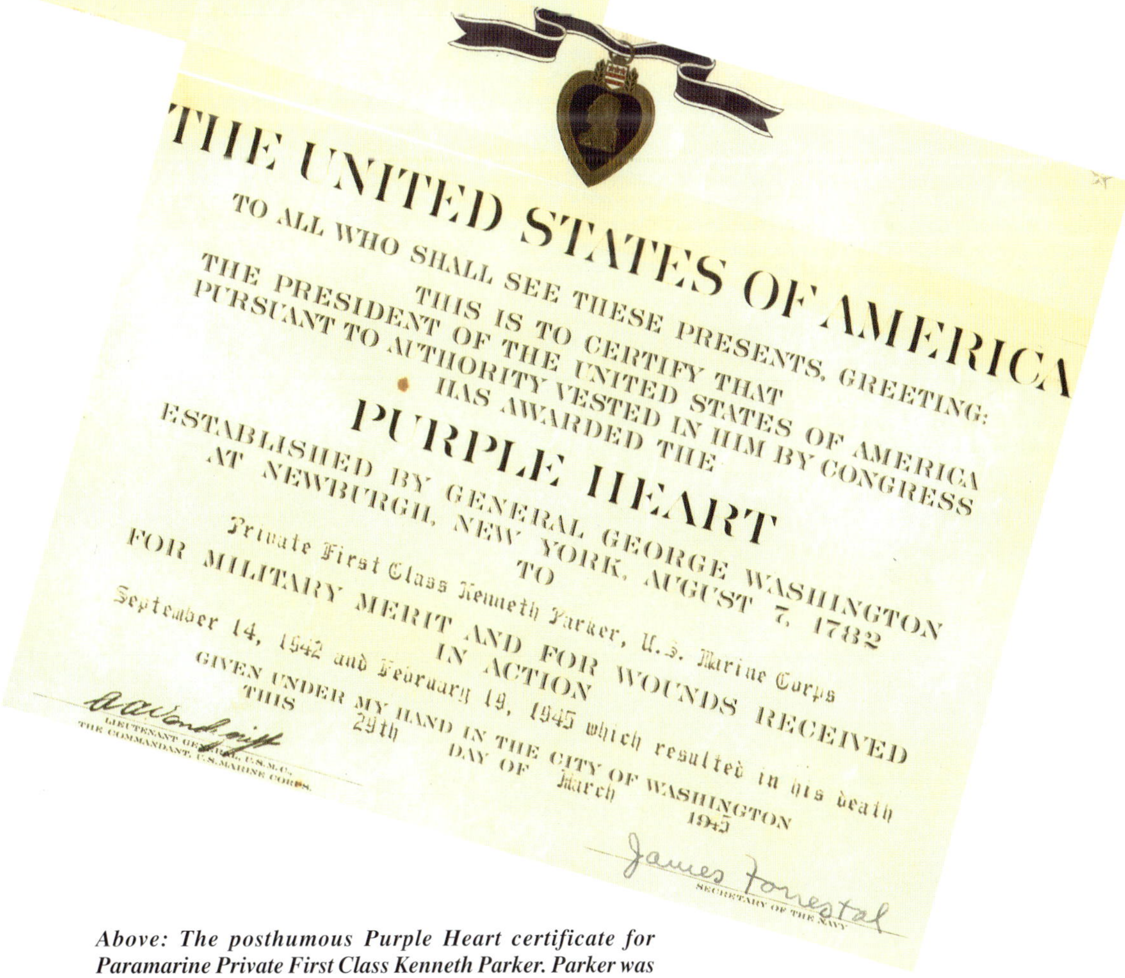

THE UNITED STATES OF AMERICA
TO ALL WHO SHALL SEE THESE PRESENTS, GREETING:
THIS IS TO CERTIFY THAT
THE PRESIDENT OF THE UNITED STATES OF AMERICA
PURSUANT TO AUTHORITY VESTED IN HIM BY CONGRESS
HAS AWARDED THE
PURPLE HEART
ESTABLISHED BY GENERAL GEORGE WASHINGTON
AT NEWBURGH, NEW YORK, AUGUST 7, 1782
TO
Private First Class Kenneth Parker, U.S. Marine Corps
FOR MILITARY MERIT AND FOR WOUNDS RECEIVED
IN ACTION
September 14, 1942 and February 19, 1945 which resulted in his death
GIVEN UNDER MY HAND IN THE CITY OF WASHINGTON
THIS 29th DAY OF March 1945

LIEUTENANT GENERAL, U.S.M.C.
THE COMMANDANT, U.S. MARINE CORPS.

James Forrestal
SECRETARY OF THE NAVY

Above: The posthumous Purple Heart certificate for Paramarine Private First Class Kenneth Parker. Parker was killed in action on Iwo Jima. (L. Jacobsen)

Left: The medals of Paramarine Private First Class Frederick H. Nicholls, Company B, 1st Marine Parachute Battalion, including his "Parachute Man" patch and Honorable Discharge pin, known as the "Ruptured Duck." (L. Jacobsen)

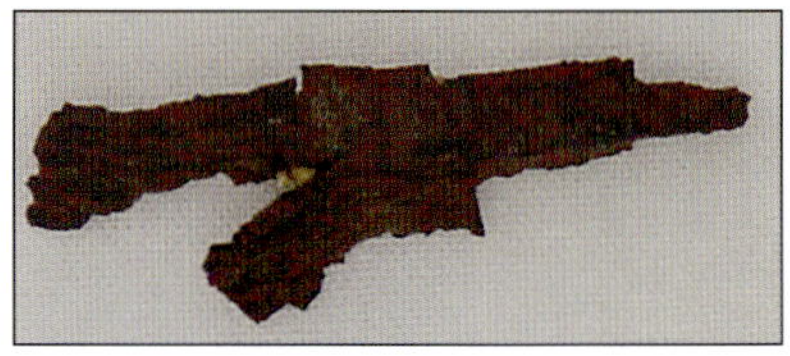

Nicholls kept the piece of shrapnel that struck him as a souvenir and kept it with his Purple Heart. (L. Jacobsen)

Below: The Purple Heart awarded to Paramarine Private First Class Frederick Nicholls for wounds received on Bougainville. (L. Jacobsen)

Below left: The reverse of Nicholl's Purple Heart showing the inscription. (L. Jacobsen)

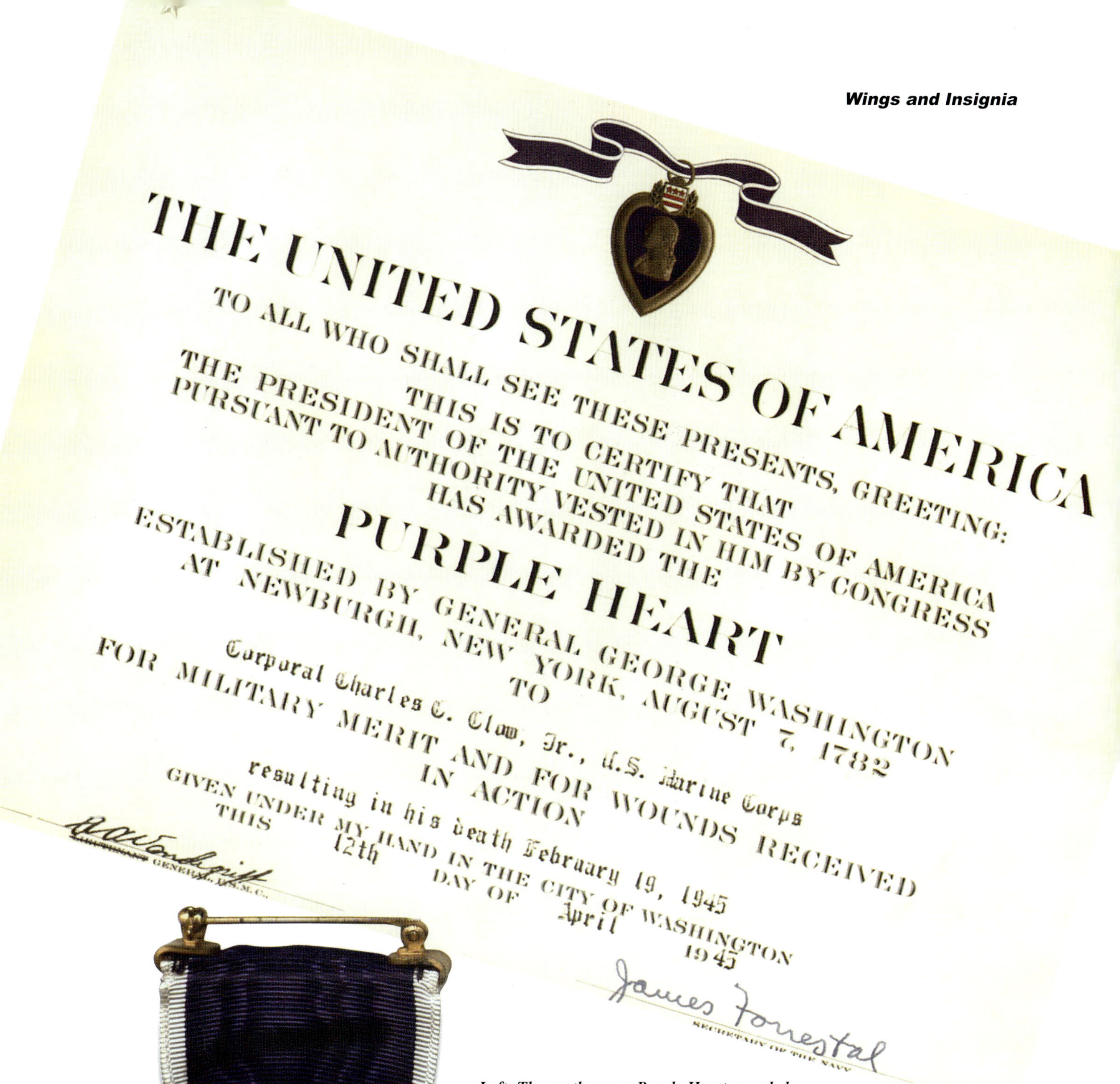

THE UNITED STATES OF AMERICA
TO ALL WHO SHALL SEE THESE PRESENTS, GREETING:
THIS IS TO CERTIFY THAT
THE PRESIDENT OF THE UNITED STATES OF AMERICA
PURSUANT TO AUTHORITY VESTED IN HIM BY CONGRESS
HAS AWARDED THE
PURPLE HEART
ESTABLISHED BY GENERAL GEORGE WASHINGTON
AT NEWBURGH, NEW YORK, AUGUST 7, 1782
TO
Corporal Charles C. Clow, Jr., U.S. Marine Corps
FOR MILITARY MERIT AND FOR WOUNDS RECEIVED
IN ACTION
resulting in his death February 19, 1945
GIVEN UNDER MY HAND IN THE CITY OF WASHINGTON
THIS 12th DAY OF April 1945

A A Vandegrift
GENERAL, U.S.M.C.

James Forrestal
SECRETARY OF THE NAVY

Left: The posthumous Purple Heart awarded to Corporal Charles C. Clow, Jr. and sent to his parents. (L. Jacobsen)

Above: The posthumous Purple Heart Certificate sent to the parents of Corporal Charles C. Clow, Jr. Military records indicate Corporal Clow died from third degree burns over his entire body on Iwo Jima. His platoon leader's handwritten letter to his Clow's parents said he "died instantly from a shell splinter above the heart as a result of a direct hit and did not suffer at all," a merciful fiction to ease the grief of loved ones at home. (L. Jacobsen)

CHAPTER SIX

Navy Personnel

The Navy played an instrumental role in the development of the parachute in the interwar years. Here a Navy Chief Petty Officer wears the Russel parachute, developed in 1926.

As this series of photographs of Navy test parachutists taken between the wars shows, the development of the military parachute took some very brave men.

Navy personnel were also instrumental in the establishment of the Paramarine program. Note the Navy enlisted rigger on the left inspecting this stick of Paramarine trainees boarding for a training jump in 1941. (E. Queen)

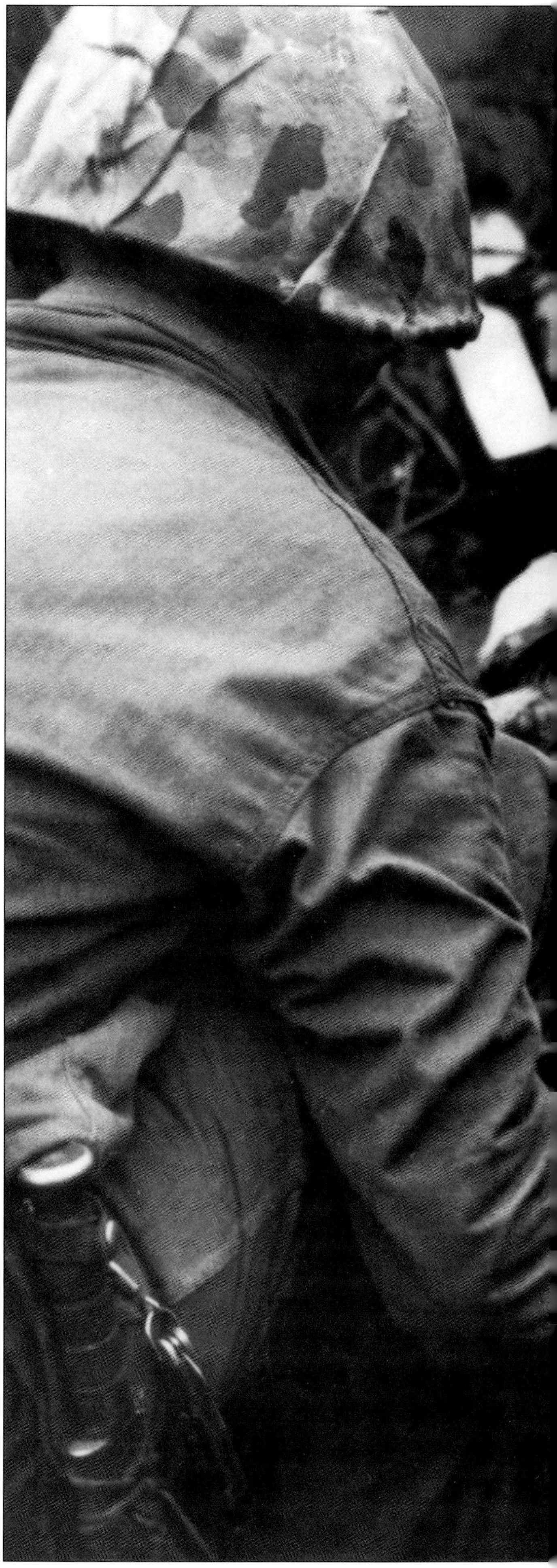

Above: Lieutenant Commander Harold Schwartz, MC, USN, wears the USMC regulation winter service green uniform with parachute wings. Schwartz served in Headquarters Company, 1st Marine Parachute Battalion, and was wounded in action on Gavutu. The Navy has always provided specialists to the Marine Corps, including medical personnel, Chaplains, and Naval Gunfire Liaison Officers (NGLOs). Navy enlisted and officer personnel assigned to the Marine Corps could then (and still may) wear Marine Corps uniforms instead of their Navy uniforms if they conform to Marine Corps grooming standards. The practice was relatively common during World War II, but is rare today. (C. Mason)

Right: Navy medical personnel operating on a badly wounded Marine on Iwo Jima from a field medical kit. The Corpsman at left wears a Kabar knife on his right hip. A "Corpsman" is, by definition, an enlisted man in a medical rating (then, a Pharmacist Mate, or PM, today a Hospitalman, or HM.) Doctors are officers commissioned in the Medical Corps (MC) of the Navy. Navy Corpsmen assigned to the Marine Corps were (and still are) affectionately and universally called "Doc." (USMC)

Inset opposite: The only difference between the HBT utility uniforms worn by Navy personnel and Marine Corps personnel was the Navy stencil over the breast pocket in lieu of the USMC stencil. (C. Mason)

Above: A variety of Navy Corpsman rank insignia for wear by Navy enlisted medical personnel assigned to the Marine Corps who wore the green winter service uniform. (C. Mason)

Right: For a period of about a year after the invasion of Tarawa, from November 1943 until October 1944, the Navy assigned radiomen to six Joint Assault Signal Companies (JASCO's) which were created to coordinate naval gunfire support during amphibious landings. Each had a strength of about 500 Marines and Sailors, about 100 of whom during that year were enlisted Navy radiomen. This rare Navy Radioman Third Class (RM3) sleeve rating is for wear by one of them on the Marine Corps khaki summer service uniform. After October 1944, enlisted Marines took over the radio billets and only a handful of Navy personnel remained. (C. Mason)

A camouflage painted M-1 steel helmet worn by a Navy Corpsman assigned to the Marine Corps in the Pacific during World War II. The rank insignia for a Navy Pharmacist's Mate Third Class (PM3) can be seen scratched in the paint on the front of the helmet. (C. Mason)

In this lighting the insignia can be seen more clearly. (C. Mason)

A Navy field medical kit. This kit came from the estate of a Navy doctor in northern Virginia. (C. Mason)

The medical kit shown open with typical contents. (C. Mason)

Paramarine Navy doctor Wendell White wears the "Kit, First Aid, Paratrooper" open for inspection on New Caladonia in April 1943. The Navy lieutenant (rank equivalent of a Marine Corps captain) inspecting the kit is wearing a regulation Navy officer's cotton long sleeve khaki shirt with parachute wings above the breast pocket. (USMC)

The paratrooper first aid kit shown open with contents. (C. Mason)

The reverse of the kit, showing the blue-gray chest straps on this particular kit. (C. Mason)

The inventory sheet for the paratrooper first aid kit in its celluloid protective sleeve. (J. Busardo)

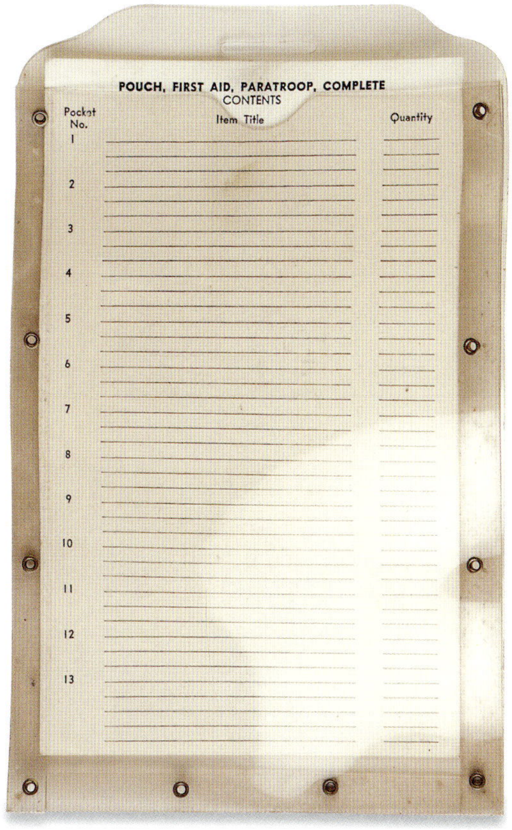

The reverse of a kit showing the bag unfolded. (C. Mason)

Below: Another paratrooper first aid kit shown without contents. This kit has white straps, a minor manufacturing variation. (J. Busardo)

A Navy Corpsman on Tarawa wears the "Kit, First Aid, Paratrooper," indicating wider issue than just to Paramarines. Interestingly, a number of other items of Paramarine gear have been spotted in photos from Tarawa, shown earlier in this book, including the first model smock (or cut down smock), jump boots, and a long, unidentified camouflage garment. The painted white discs on the helmets and uniforms identified Navy medical personnel. Rank details were sometimes hand written or painted into the discs. (USMC)

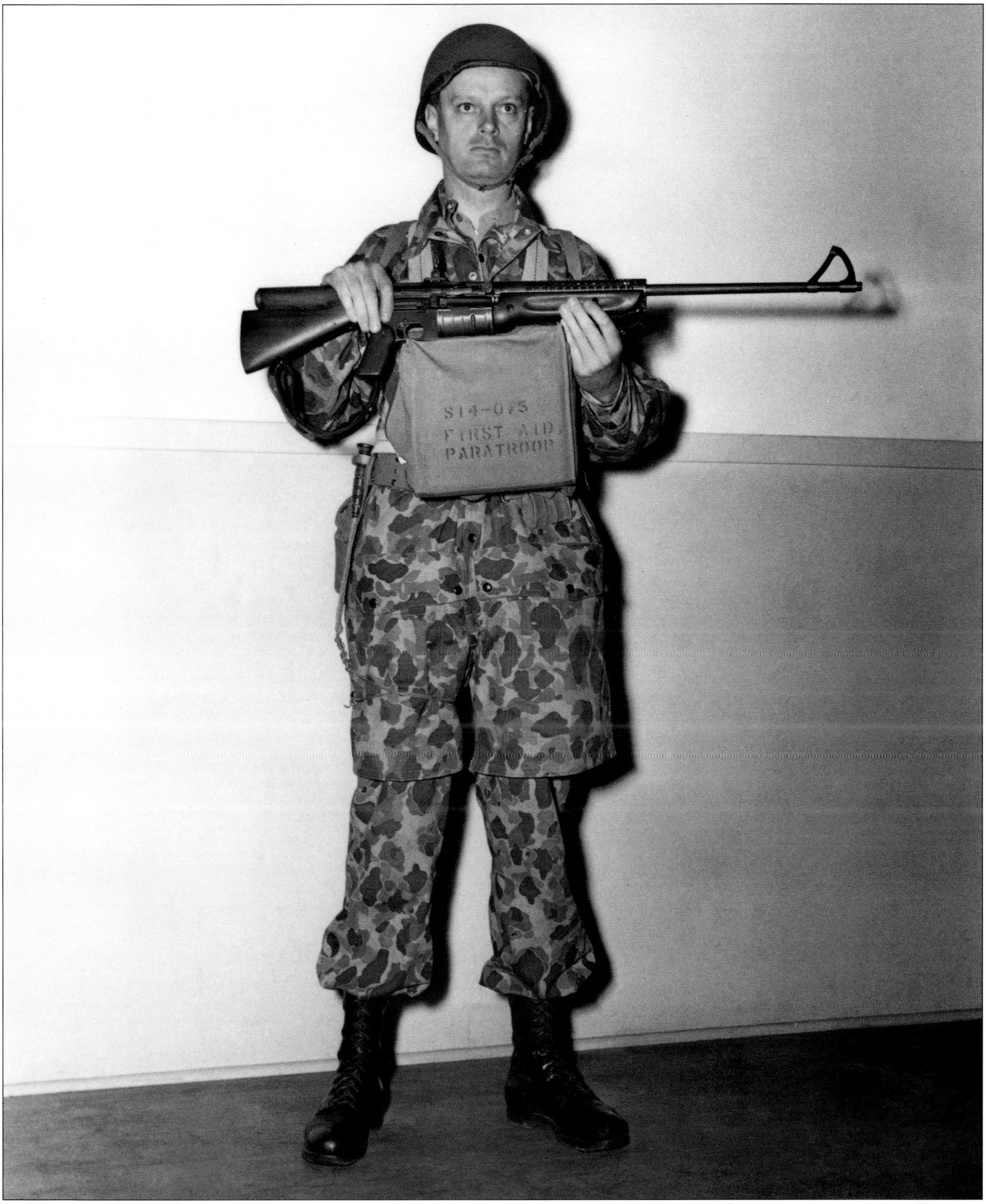

Opposite
Top: Another photo from Tarawa also shows the white disc identification system in use. (USMC)

Bottom: In addition to officer and enlisted medical personnel, the Navy also provides for the religious needs of the Marine Corps. Chaplains of various denominations were (and are) commissioned Navy officers, and during World War II were often affectionately addressed by Marines as "Padre" (a term still heard among Navy officers today.) Here, a Navy Chaplain, Father Joseph P. Mannion, at parachute training in May 1943. Father Mannion completed his training, earned his silver wings, and served in the 4th Marine Parachute Battalion. (USMC)

Above: Lieutenant Robert J. Leonard, USN, Company K, 3rd Marine Parachute Battalion, holds Major Torgerson's one-of-a-kind Johnson carbine. He is wearing a medical kit on his chest stenciled "S14-075 First Aid Paratrooper" at Camp Elliot in February 1943, a second model smock and a Kabar knife. (USMC)

CHAPTER SEVEN

Personal Items

The sea chest of Paramarine Sergeant Harold A. Hoeschen, Company C, 1st Marine Parachute Battalion. (L. Jacobsen)

Sergeant Hoeschen's chest with lid closed, showing the standard hardware. (L. Jacobsen)

Sergeant Hoeschen stenciled his name on the lid in accordance with USMC regulations. (L. Jacobsen)

This parachute carrying bag served as a seabag for Paramarine Denton D.Howard. Howard was wounded on Iwo Jima with 27th Marines. (C. Mason)

Howard's name stenciled on the top of the bag. In Marine Corps nomenclature "1st Marine Paratroops" meant "1st Marine Parachute Regiment." (C. Mason)

The parachute carrying bag used by Paramarine Sergeant Hoeschen. His stenciled name can be seen on the left of the bag. (L. Jacobsen)

Left: The staple diet of a Paramarine in the field usually consisted of K Rations, as often as not. When in Australia, the steady diet of mutton made many vow never to eat lamb again. (C. Mason)

The Chesterfield cigarettes came in the K-Rations. The World War II military issue Lucky Strikes lost their traditional green packet during the war so the green dye could be used for military purposes. Photos of GI icon Betty Grable and other popular movie actresses came in some brands of cigarettes as a promotion. (C. Mason)

The USMC issue folding 3-legged cooker and USMC issue heating tabs. Both were lightweight and took up very little space. (C. Mason)

Above: A typical Marine's "housewife," a small sewing kit for sewing buttons back on, darning socks and other emergency repairs in the field. Right: Typical contents of a "housewife." (C. Mason)

Left: A more elaborate sewing kit, which might be kept in a footlocker or seabag in the rear. (C. Mason)

Below left: der. Skin rashes and fungal infections were a constant torment in tropical areas like Guadalcanal and Bougainville. (C. Mason)

Below: Boot polish was a necessary item in the foot locker during training for the gear and uniform inspections which preceded "Liberty Call,." the Navy and Marine Corps term for time off duty. (C. Mason)

Bottom right: Wartime playing cards and wartime Sheik condoms. Long ocean transits provided ample time to use the former, but most Marines found opportunity to use the latter only for waterproofing items of equipment. The Breakers Club in Hawaii was a Navy Recreation Center during the war. (C. Mason)

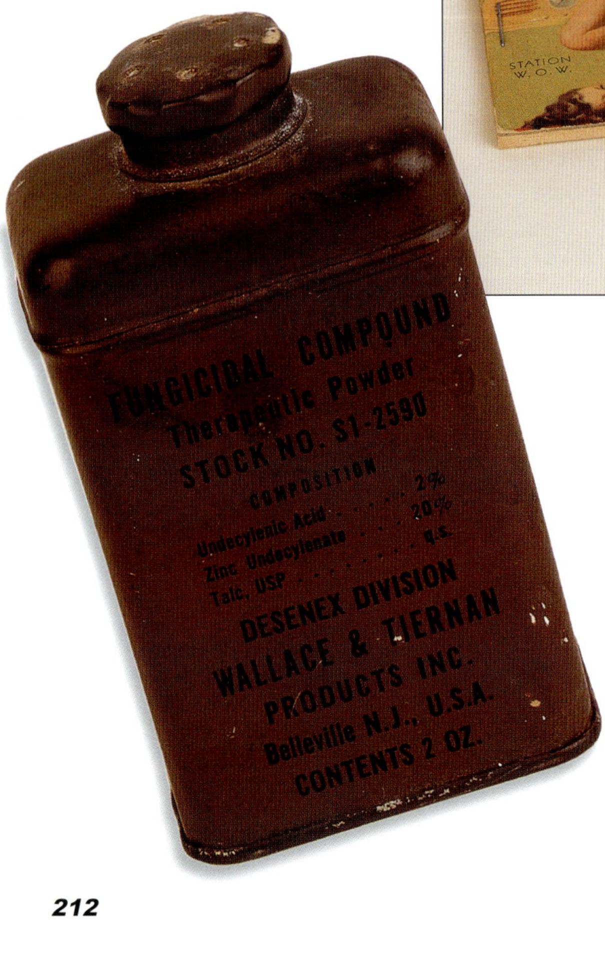

Right: Corporal (here a PFC) Charles C. Clow, Jr. (center) and two buddies clown for a photo on liberty in the "Tiajuana Jail." Clow's gold parachute wings are just visible in the photograph. Careful review of Clow's service record shows he was not trained or qualified as a Rigger at any time during the war, thus apparently also wore the "gold wings" out of personal preference. The fact that he's wearing the gold wings out on liberty (not just for the private studio photograph shown earlier) indicates at least tacit acceptance of this practice by his officers, who might easily spot him out in town with them on. (L. Jacobsen)

Below: While in a combat zone, Marines packed light. A simple "ditty bag" made of sturdy HBT material with a drawstring might contain a razor, some issue blades, a metal mirror and some soap – if any could be scrounged. (C. Mason)

In the rear areas, more elaborate grooming kits like this one might be stored in the footlocker or seabag. They contained all the necessary elements to get "spiffed up" for a night on the town, including after shave lotion, or "foofoo juice" in Marine parlance. (C. Mason)

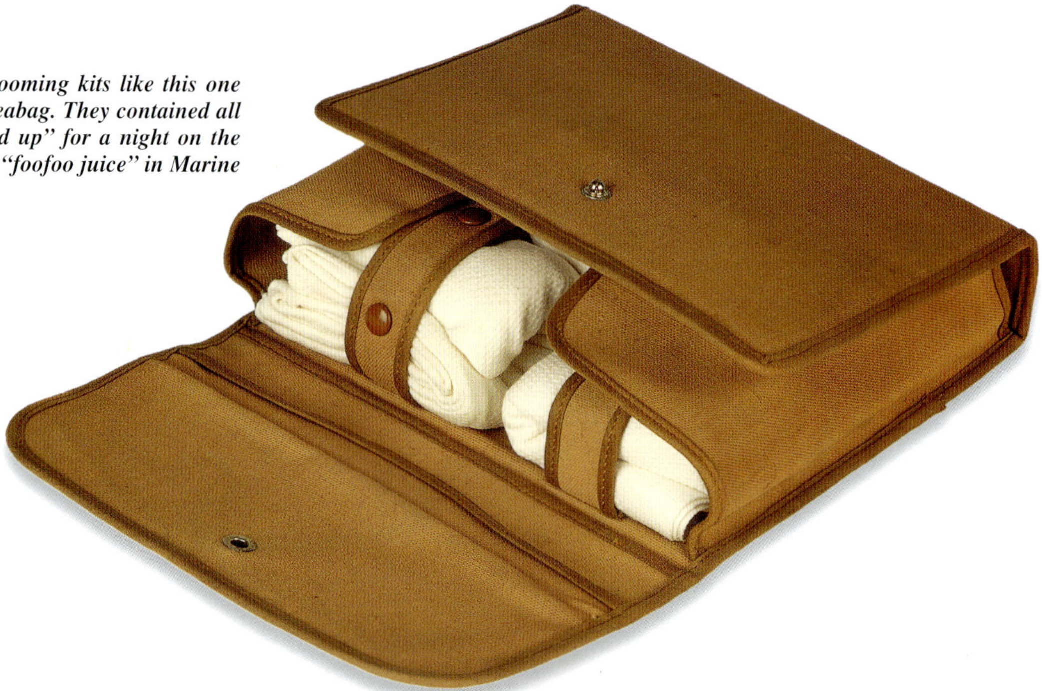

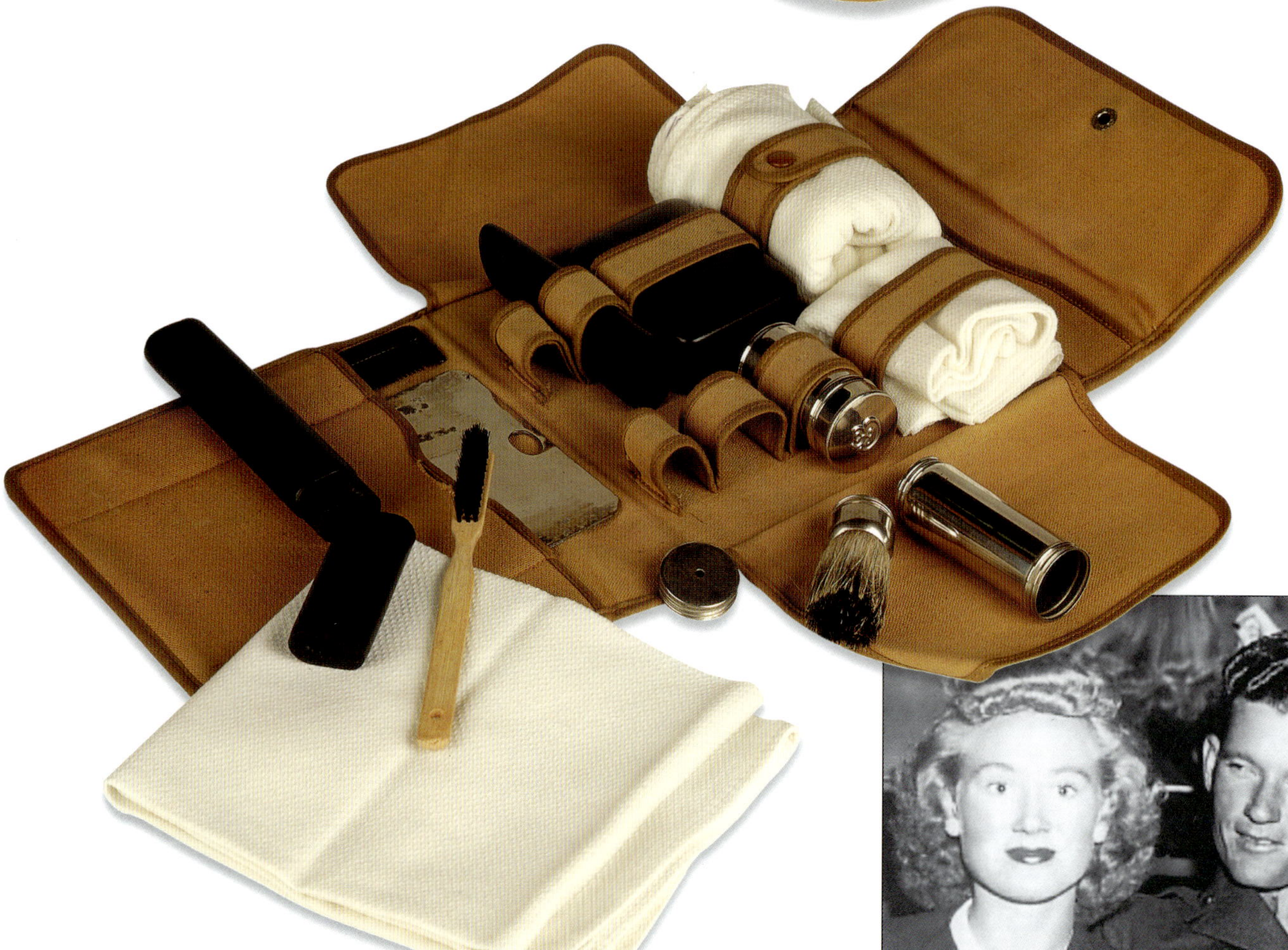

A typical grooming kit for the rear area opened to show the contents. (C.Mason)

Right: A night on the town on Liberty. The photo is labeled "Paris Inn 1944. Corporal Hermann C. Glass is the Paramarine on the right. (E. Queen)

A common tie bar for the "field scarf" of the enlisted uniform. (L. Jacobsen)

This bracelet, with jump wings affixed, belonged to Paramarine Sergeant Edward G. Gruber. Gruber was wounded in action on Iwo Jima with 27th Marines. It is engraved "1st Chutes USMC." Engraved on the other side of the bracelet is "Ed Gruber." (L. Jacobsen)

Right: Gruber in his winter service uniform after basic training at Parris Island. (L. Jacobsen)

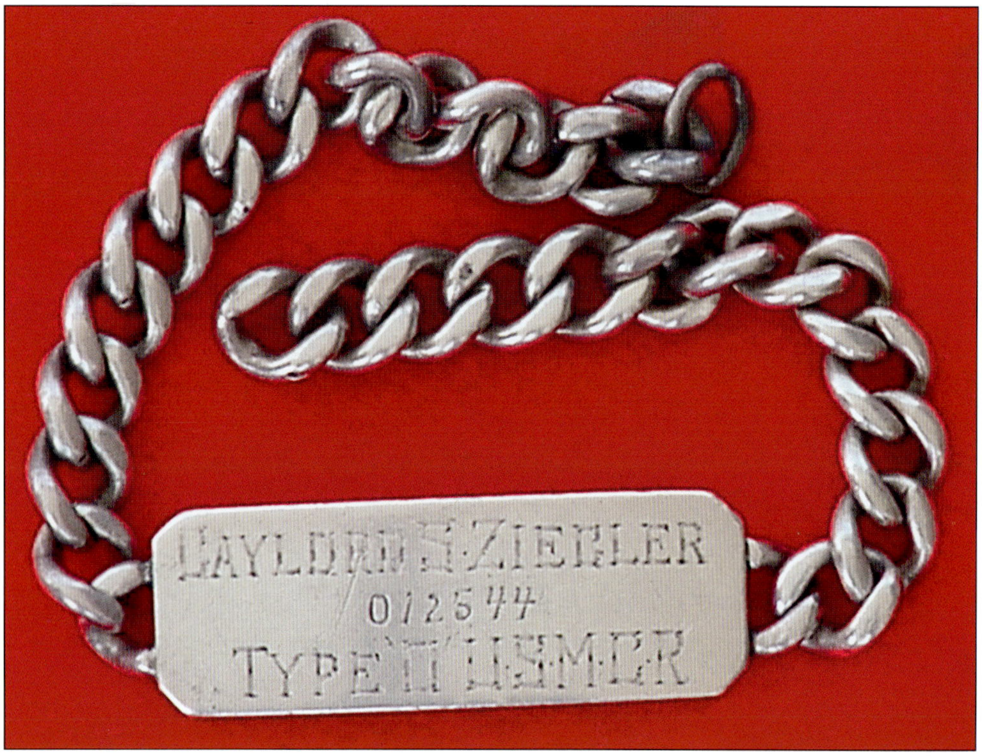

Above: This bracelet belonged to Paramarine officer Gaylord Ziegler, who was assigned to Headquarters Company, 3rd Marine Parachute Battalion. Ziegler was wounded on Iwo Jima with 28th Marines. Right: The reverse of Ziegler's bracelet bears an inscription from his wife, Barbara. (L. Jacobsen)

Above: First Lieutenant Ziegler's dogtag issued in July 1941. Right: Ziegler's dogtag issued in April 1943 omits his rank but includes his serial number, which is also seen on his bracelet. (L. Jacobsen)

This interesting matchbook cover is one of the few surviving relics of the Marine Corps assault glider program, which arose simultaneously with the Paramarine program as a result of the USMC's interest in the German invasion of Crete. (C. Mason)

The front of the matchbook under the striking pad shows the final location of the glider program, Eagle Mountain Lake, Texas. The glider program was terminated in June 1943, and the few gliders on hand were turned over to the Army. (C. Mason)

USMC gliders very early in the program during World War II at Parris Island. The marking "US Marines" can be seen on the fuselages of both aircraft. (USMC)

Above: Typical pin up art matchbooks of the period. (C. Mason)

Left top: A matchbook from the Marine training camp at New River, North Carolina. Originally named simply Marine Corps Base New River, it was rechristened in honor of former Marine Corps Commandant Lieutenant General John Archer Lejeune. One of the Marine Corps jump schools was located here. (C. Mason)

Left: The front of the matchbook bears the world famous Eagle, Globe and Anchor (EGA) emblem of the Marine Corps. (C. Mason)

The Marine Corps ID Card of Paramarine Corporal Charles C. Clow, Jr. (L. Jacobsen)

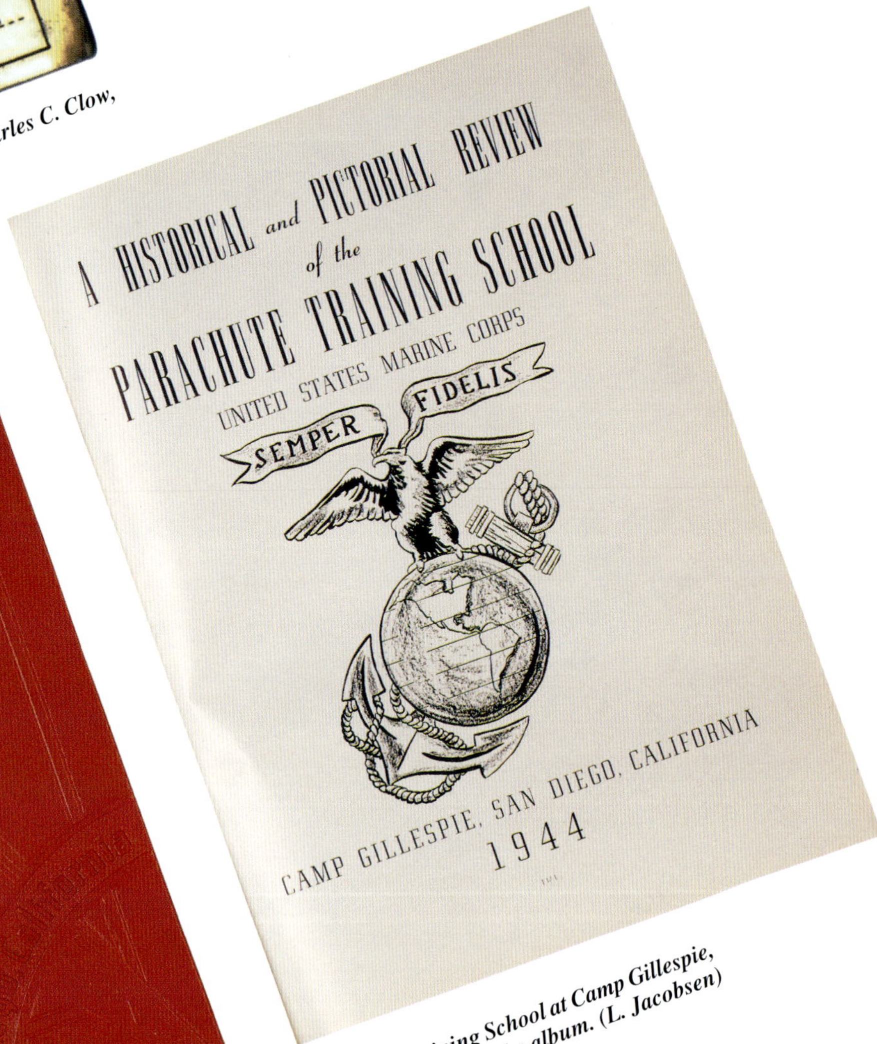

This keepsake photo album, or yearbook, of the USMC Parachute Training School at Camp Gillespie, California, was published early in 1944. At right is one of the title pages of the album. (L. Jacobsen)

Right: This particular album was mailed home in this box by Paramarine Milton Peterson of the 4th Marine Parachute Battalion. Peterson was killed in action on Iwo Jima with 27th Marines. Left: Paramarine Milton Peterson. (L. Jacobsen)

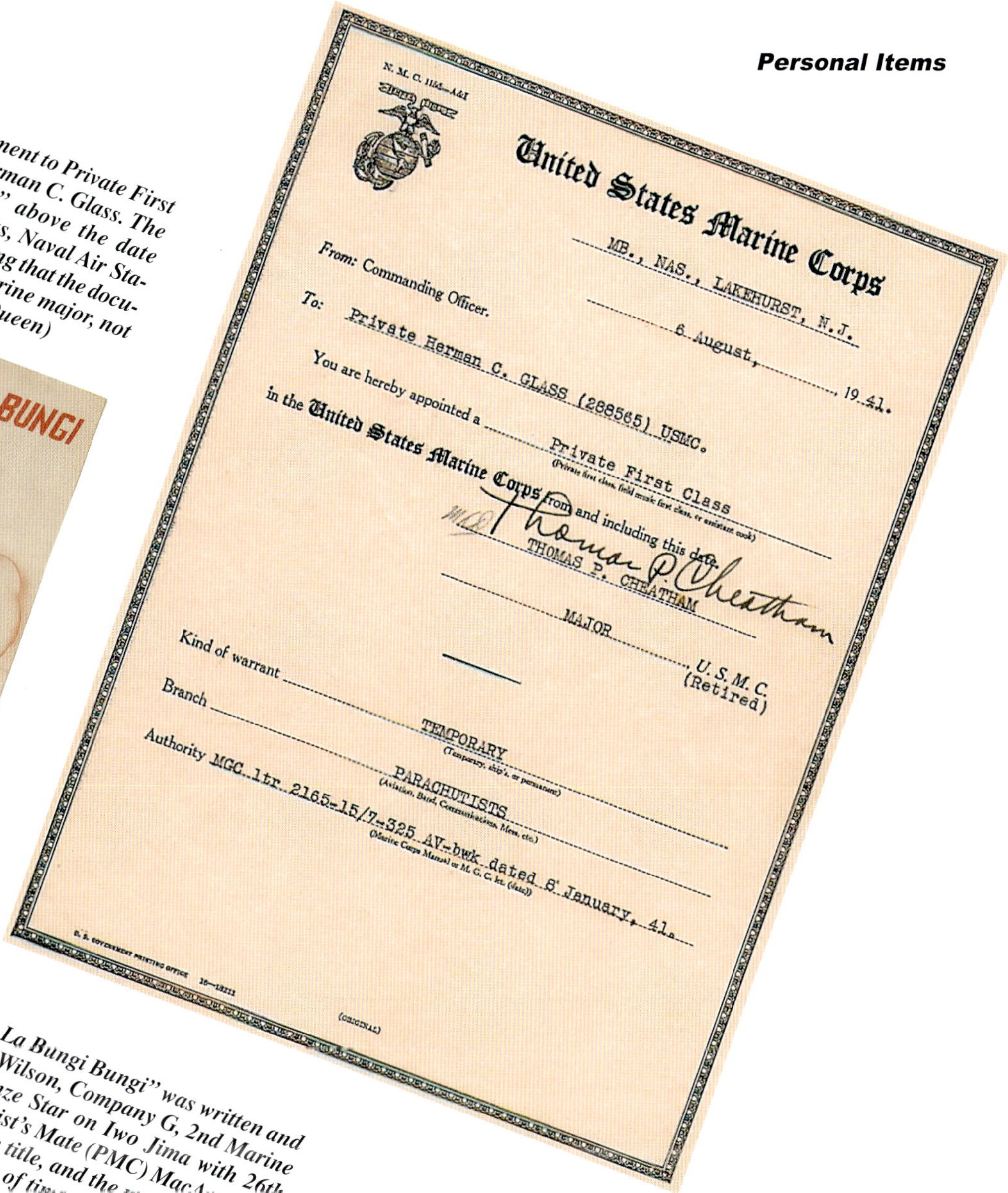
N. M. C. 1165—A&I

United States Marine Corps

MB., NAS., LAKEHURST, N.J.

From: Commanding Officer.

6 August, 1941.

To: Private Herman C. GLASS (288565) USMC.

You are hereby appointed a Private First Class (Private first class, field music first class, or assistant cook)

in the United States Marine Corps from and including this date.

Thomas P. Cheatham
THOMAS P. CHEATHAM
MAJOR
U.S.M.C. (Retired)

Kind of warrant ——

Branch TEMPORARY (Temporary, ship's, or permanent)

PARACHUTISTS (Aviation, Band, Communications, Mess, etc.)

Authority MGC. ltr. 2165-15/7-325 AV-bwk dated 6 January, 41. (Marine Corps Manual or M. C. C. ltr. (date))

U. S. GOVERNMENT PRINTING OFFICE 16—18233

(ORIGINAL)

Right: A promotion document to Private First Class (later Corporal) Herman C. Glass. The abbreviation "MB., NAS" above the date stands for "Marine Barracks, Naval Air Station" Lakehurst. It is interesting that the document is signed by a retired Marine major, not an officer on active duty. (E. Queen)

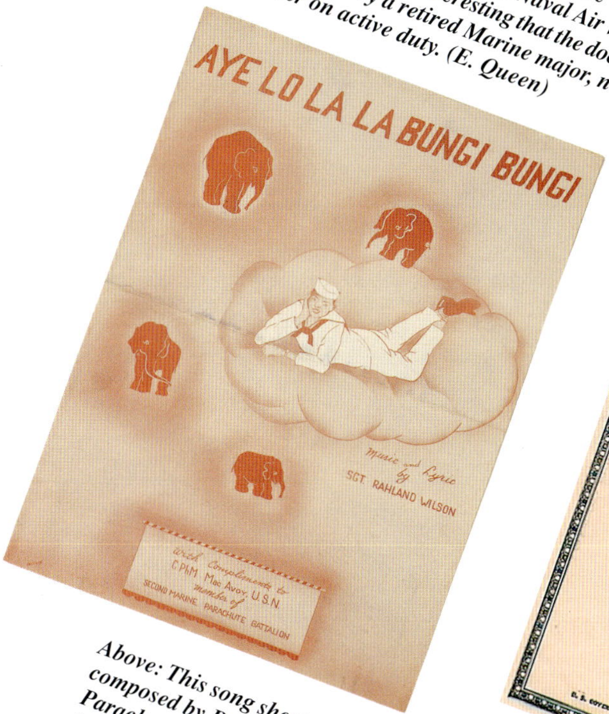

Above: This song sheet for the song "Aye Lo La Bungi Bungi" was written and composed by Paramarine Sergeant Rahland Wilson, Company G, 2nd Marine Parachute Battalion. Wilson earned the Bronze Star on Iwo Jima with 26th Marines. Note the dedication to Chief Pharmacist's Mate (PMC) MacAvoy also of 2nd Battalion. The meaning of the words, the title, and the motif of a sailor surrounded by elephants may be lost in the mists of time. (C. Mason)

Left: The pocket guide to Australia given to Sailors and Marines going ashore or garrisoned in Australia. Right: The companion guide to New Zealand. (C. Mason)

Above: This postcard from Camp Lejeune depicts "Paramarines in Action." (C. Mason) Right: This postcard depicts the chapel at Naval Air Station Lakehurst, home of the Navy Parachute Materiel School Lakehurst, one of the USMC jump schools. NAS Lakehurst is also remembered as the site of the Hindenburg Disaster in 1937, when the hydrogen-filled German dirigible exploded in a huge fireball with a loss of 36 lives. (C. Mason)

Above: This postcard depicts "Parachute Troops" of the Marine Corps. (C. Mason) Right: A children's bubble gum trading card of World War II depicts "Marine Parachute Troops" jumping from a blimp. In fact, early in the parachute program, Marine Corps parachutists did jump from such airships, which were stationed at Lakehurst for anti-submarine patrols over the Atlantic. Because they could remain on station for such long periods, blimps were actually very useful in such patrols. (C. Mason)

Below: A souvenir photo from training at Lakehurst in 1941. This photograph belonged to Corporal Hermann C. Glass, 1st Marine Parachute Regiment. (E. Queen)

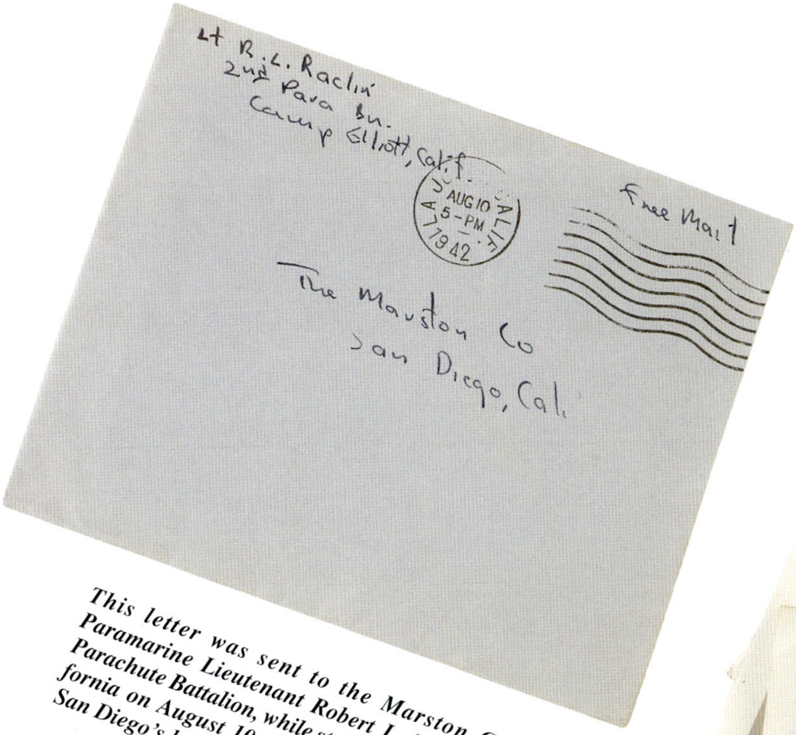

This letter was sent to the Marston Company by Paramarine Lieutenant Robert L. Raclin, 2nd Marine Parachute Battalion, while stationed at Camp Elliott, California on August 10, 1942. The Marston Company was San Diego's leading department store. (L. Jacobsen)

Right: Typical photographs taken as souvenirs of jump school. These belonged to Paramarine Edward L. McClure, Company F, 2nd Marine Parachute Battalion. (L. Jacobsen)

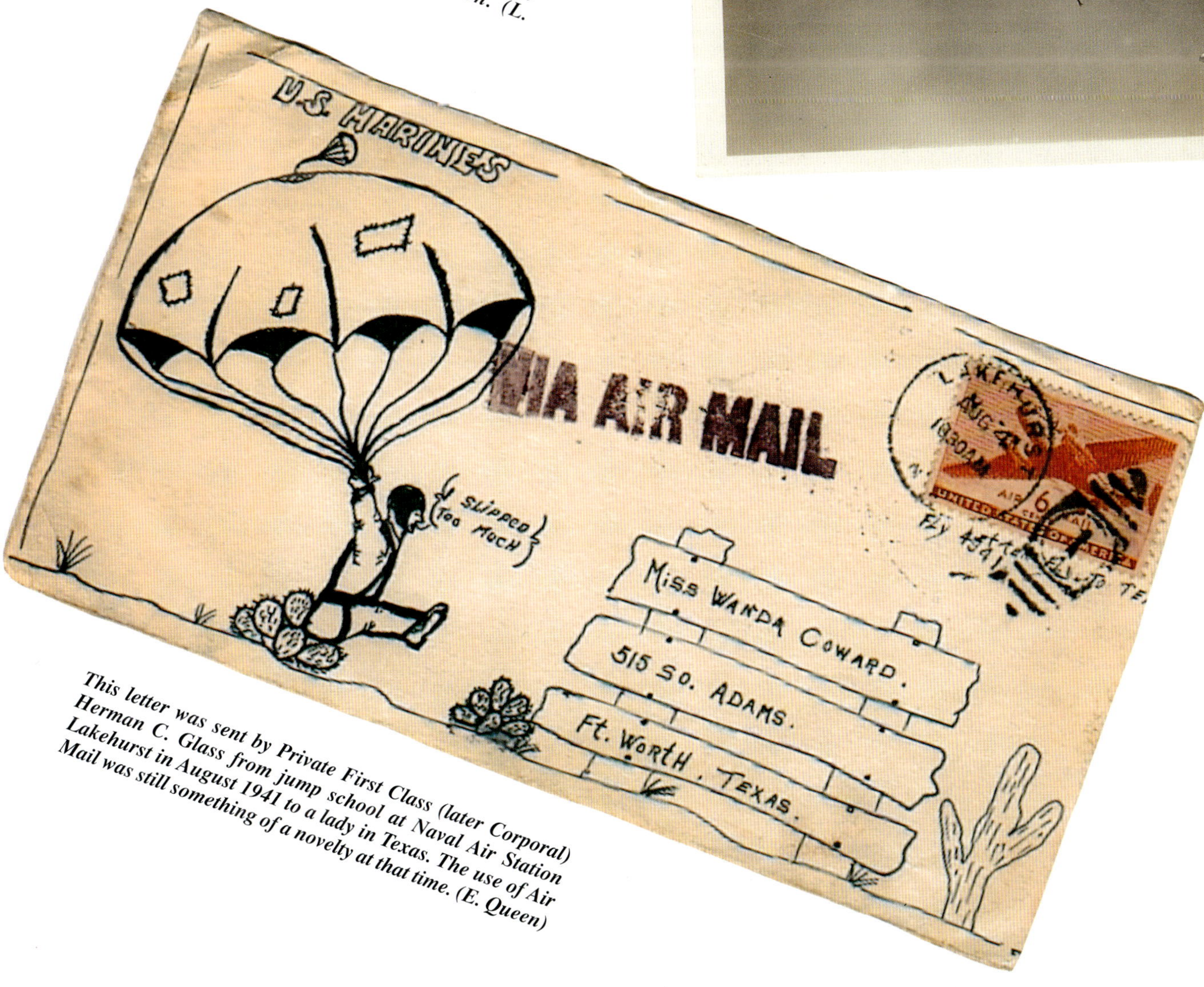

This letter was sent by Private First Class (later Corporal) Herman C. Glass from jump school at Naval Air Station Lakehurst in August 1941 to a lady in Texas. The use of Air Mail was still something of a novelty at that time. (E. Queen)

The photo album of Paramarine Sergeant Edward G. Gruber. (L. Jacobsen)

Above left: This photograph from Gruber's album shows wounded being loaded aboard a Higgins Boat somewhere in the Pacific. Center: This photograph, apparently taken by Gruber, shows Paramarines waiting to go ashore, probably in Bougainville. Right: This photograph from Gruber's album shows an unidentified Paramarine on board ship. Note the column of smoke on the shore and the elastic foliage band over the Marine's helmet cover. (L. Jacobsen)

Right: Typically, Gruber's album contains several stock photos of topless native girls in the South Pacific, a fantasy inspired by sailor's tales of Tahiti. In reality, few Marines ever saw a native woman. (L. Jacobsen)

Below: Also typical are these photographs of dead Japanese in Gruber's photo album. Unlike the native girls, most combat Marines saw such terrible sights in large numbers. (L. Jacobsen)

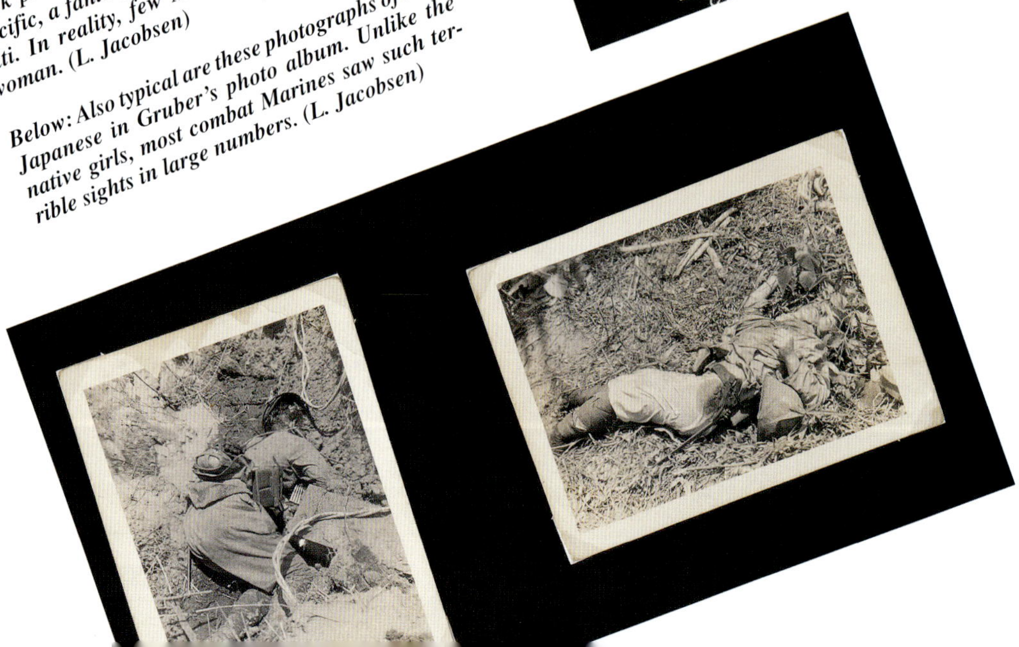

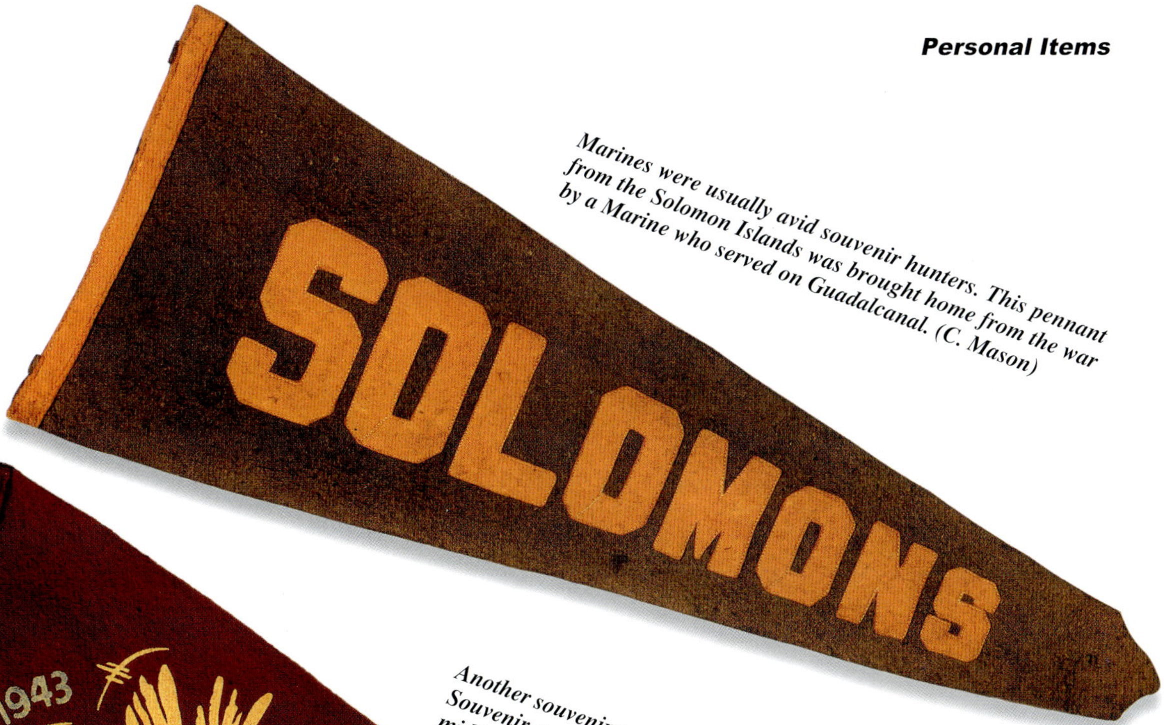

Marines were usually avid souvenir hunters. This pennant from the Solomon Islands was brought home from the war by a Marine who served on Guadalcanal. (C. Mason)

Another souvenir pennant, this one from Australia, circa 1943. Souvenir pennants were very popular with Americans in the middle of the 20th Century. (C. Mason)

During the war, the pin-up postcards of Zöe Mozert were also popular with the troops. (C. Mason)

A typical Mozert postcard depicts a leggy female parachutist. (C. Mason)

Right: Two Marines and a Navy Corpsman (center) examine Japanese currency found on Roi island in the Marshalls. (USMC)

Below: Typical souvenirs brought home by Marines taken from dead Japanese included flags, insignia, currency, swords and the prized Nambu pistol. The currency and insignia in this photo were sent home by Paramarine Milton Peterson before he was killed on Iwo Jima. (J. Busardo, G. Petersen, L. Jacobsen)

Below: Marines somewhere in the Pacific examine an abandoned Japanese hut for souvenirs such as the "meatball" flag, left. (USMC)

Right: Other Marines on Namur display a prized samurai sword and Arisaka rifles while sampling stocks of captured Japanese rice beer found on the island. Note that the man in the background at the top of the hill is carrying a combat shotgun with mounted bayonet. (USMC)

Below: Marines on Namur (Kwajalein Atoll) display a small flag and currency souvenirs found on the island. (USMC)

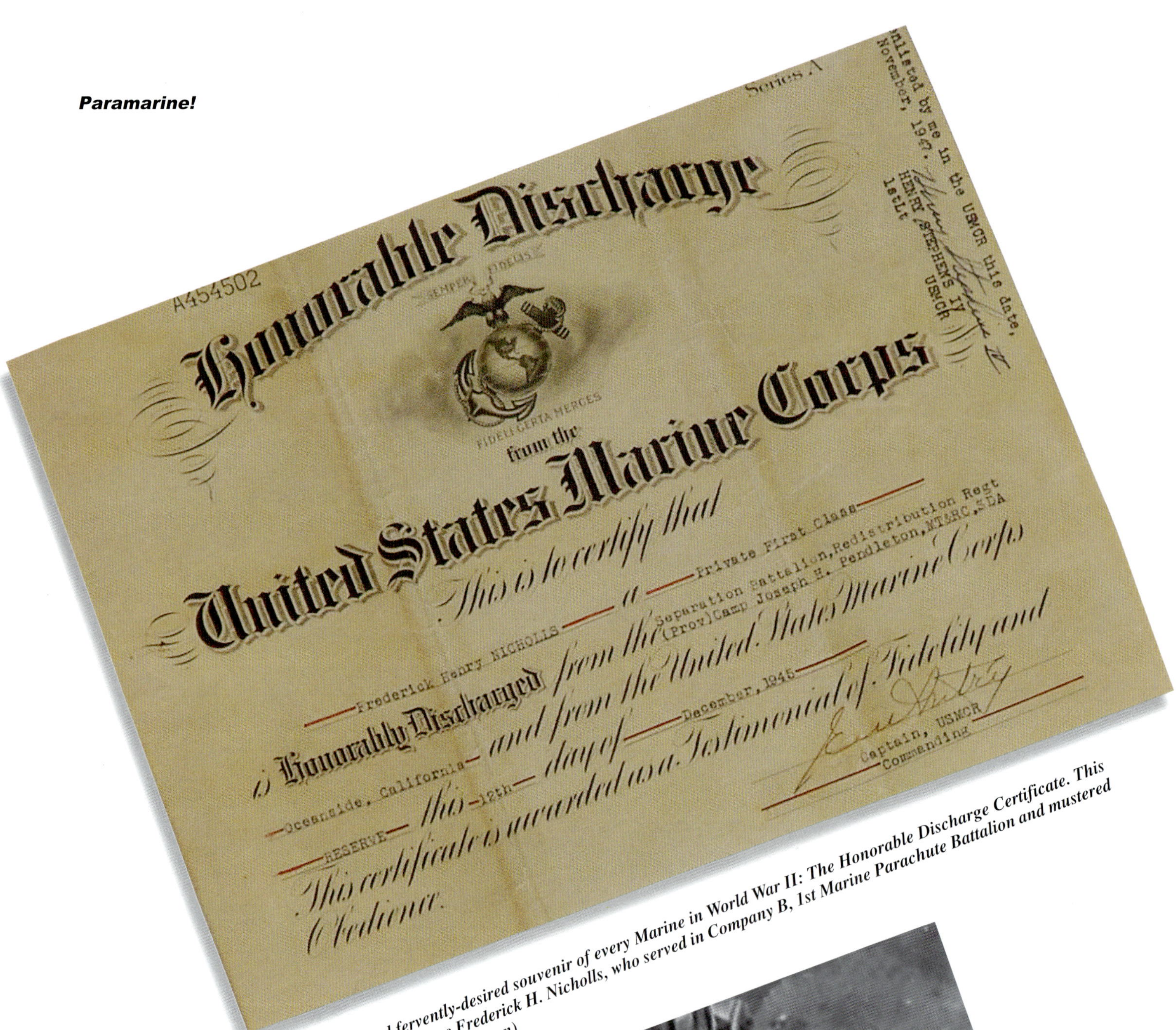
Series A

A454502

Honorable Discharge

SEMPER FIDELIS

FIDELI CERTA MERCES

from the United States Marine Corps

This is to certify that

Frederick Henry NICHOLLS a Private First Class

is Honorably Discharged from the Separation Battalion, Redistribution Regt (Prov) Camp Joseph H. Pendleton, MT&RC, SDA

Oceanside, California and from the United States Marine Corps

RESERVE this 12th day of December, 1945

This certificate is awarded as a Testimonial of Fidelity and Obedience.

Captain, USMCR
Commanding

Enlisted by me in the USMCR this date, November, 19
HENRY STEPHENS IV
1stLt USMCR

The most prized and fervently-desired souvenir of every Marine in World War II: The Honorable Discharge Certificate. This one belonged to Paramarine Frederick H. Nicholls, who served in Company B, 1st Marine Parachute Battalion and mustered out in December 1945. (L. Jacobsen)

Civvy Street. Hermann Glass after the war, sitting on the running board of his car. He is still wearing an Australian battledress jacket with his embroidered gold parachutist wings over the left breast pocket, the First Marine Division patch on his right shoulder, and an unknown patch on his left shoulder. Like most former Paramarines, Glass was clearly proud to have been a paratrooper.

Above: The fate of 407,316 young American Marines, Sailors, Soldiers, Airmen, and Coast Guardsmen who gave their lives in service of their country during World War II. A photograph of the grave marker of Paramarine Corporal Charles C. Clow, Jr, on Iwo Jima. Of the 25 individual Paramarines represented by name in this book, more than half were killed or wounded on Iwo Jima, testimony to the horrendous lethality of the epic victory on the island. (Four others of the 25 were wounded on Gavutu or Bougainville, a 65 percent casualty rate among this sample group). (L. Jacobsen)

Below: The Fifth Marine Division cemetery on Iwo Jima. Corporal Clow's grave is in the foreground and Mount Suribachi stands in the background. The Marine Corps comprised just five percent of US uniformed forces during World War II but suffered ten percent of our casualties. Americans should never forget the sacrifice these young men made to preserve our freedom.